TRENDS IN
BIOLOGICAL ANTHROPOLOGY

VOLUME 2

Edited by

MALIN HOLST AND MICHELLE ALEXANDER

Proceedings of the British Association for
Biological Anthropology and Osteoarchaeology
15th Annual Conference in
York (13th to 15th September 2013)

Monograph Series Volume 2

Volume editors: Malin Holst
and Michelle Alexander (University of York)

Monograph Series editor: Tina Jakob (Durham University)

OXBOW | books
Oxford & Philadelphia

Published in the United Kingdom in 2018 by
OXBOW BOOKS
The Old Music Hall, 106–108 Cowley Road, Oxford OX4 1JE

and in the United States by
OXBOW BOOKS
1950 Lawrence Road, Havertown, PA 19083

© Oxbow Books and the individual contributors 2018

Paperback Edition: ISBN 978-1-78570-620-2
Digital Edition: ISBN 978-1-78570-621-9 (epub)

A CIP record for this book is available from the British Library

Library of Congress Control Number: 2018947536

Typeset in India by Versatile PreMedia Services. www.versatilepremedia.com

For a complete list of Oxbow titles, please contact:

UNITED KINGDOM
Oxbow Books
Telephone (01865) 241249, Fax (01865) 794449
Email: oxbow@oxbowbooks.com
www.oxbowbooks.com

UNITED STATES OF AMERICA
Oxbow Books
Telephone (800) 791-9354, Fax (610) 853-9146
Email: queries@casemateacademic.com
www.casemateacademic.com/oxbow

Oxbow Books is part of the Casemate Group

Contents

List of Contributors

GURDYAL S. BESRA
Institute of Microbiology and Infection
School of Biosciences, University of Birmingham
Edgbaston, Birmingham
B15 2TT, United Kingdom
Email: g.besra@bham.ac.uk

JENNIE BRADBURY
Department of Archaeology, Durham University
South Road, Durham
DH1 3LE, United Kingdom
Email: jennie.bradbury@arch.ox.ac.uk

CARLA L. BURRELL
Chapel Archaeology CIC
143 Hough Green, Chester
CH4 8JR, United Kingdom
Email: c.burrell@chapelarchaeology.co.uk

RAYMOND J. CARPENTER
Chapel Archaeology CIC
143 Hough Green, Chester
CH4 8JR, United Kingdom
Email: ray@chapelarchaeology.co.uk

ANDREW T. CHAMBERLAIN
Faculty of Life Sciences, The University of Manchester
3.614 Stopford Building, Oxford Road, Manchester
M13 9PT, United Kingdom
Email: andrew.chamberlain@manchester.ac.uk

JACQUELINE Z.-M. CHAN
Division of Microbiology & Infection
University of Warwick, Coventry
CV4 7AL, United Kingdom
Email: jzm_chan@hotmail.com

CAROLE A.L. DAVENPORT
Research Centre for Evolutionary Anthropology &
Palaeoecology, School of Natural Sciences & Psychology
Liverpool John Moores University, Byrom Street, Liverpool
L3 3AF, United Kingdom
Email: c.a.turner@ljmu.ac.uk

DOUGLAS DAVIES
Durham University
South Road, Durham
DH1 3LE, United Kingdom
Email: douglas.davies@durham.ac.uk

HEIDI DAWSON-HOBBIS
Department of Archaeology and Anthropology
University of Winchester
Medecroft Building, Sparkford Road, Winchester
SO22 4NR, United Kingdom
Email: heidi.dawson-hobbis@winchester.ac.uk

HELEN D. DONOGHUE
Centre for Clinical Microbiology
Division of Infection and Immunity
Royal Free Campus, University College London, London
NW3 2PF, United Kingdom
Email: h.donoghue@ucl.ac.uk

JEAN-MICHEL DUGOUJON
Laboratoire d'Anthropologie Moléculaire et Imagerie de
Synthèse, CNRS UMR-5288 AMIS, Faculté de Médecine
Université Toulouse III Paul Sabatier
37, allées Jules Guesde
31000 Toulouse, France
Email: jean-michel.dugoujon@univ-tlse3.fr

CESAR FORTES-LIMA
Laboratoire d'Anthropologie Moléculaire et Imagerie de
Synthèse
CNRS UMR-5288 AMIS
Faculté de Médecine, Université Toulouse III Paul Sabatier
37, allées Jules Guesde
31000 Toulouse, France
Email: cesar@eurotast.eu

CHARLOTTE Y. HENDERSON
Research Centre for Anthropology and Health
University of Coimbra
Portugal
Email: c.y.henderson@uc.pt

MANDY JAY
Department of Archaeology, Durham University
South Road, Durham
DH1 3LE, United Kingdom
Email: mandy.jay@durham.ac.uk

OONA Y.-C. LEE
Institute of Microbiology and Infection
School of Biosciences, University of Birmingham
Edgbaston, Birmingham
B15 2TT, United Kingdom
Email: o.y.lee@bham.ac.uk

LAUREN MCINTYRE
Oxford Archaeology, Janus House
Osney Mead, Oxford
OX2 0ES, United Kingdom
Email: lauren.mcintyre@oxfordarch.co.uk

ANDREW MILLARD
Department of Archaeology, Durham University
South Road, Durham
DH1 3LE, United Kingdom
Email: a.r.millard@durham.ac.uk

DAVID E. MINNIKIN
Institute of Microbiology and Infection
School of Biosciences, University of Birmingham
Edgbaston, Birmingham
B15 2TT, United Kingdom
Email: d.e.minnikin@bham.ac.uk

JAMES C. OHMAN
Research Centre for Evolutionary Anthropology &
Palaeoecology
School of Natural Sciences & Psychology
Liverpool John Moores University
Byrom Street, Liverpool
L3 3AF, United Kingdom
Email: j.cohman@ljmu.ac.uk

DON O'MEARA
Science Advisor North East and Hadrian's Wall
Research Group, Historic England
Bessie Surtees House
41-44 Sandhill, Newcastle upon Tyne
NE1 3JF, United Kingdom
Email: Don.O'Meara@HistoricEngland.org.uk

MARK J. PALLEN
Quadram Institute, Norwich Research Park
Norwich Norfolk
NR4 7UA, United Kingdom
Email: mark.pallen@quadram.ac.uk

IDILKO PAP
Department of Anthropology
Hungarian Natural History Museum
Budapest, Hungary
Email: pap.ildiko@nhmus.hu

GRAHAM PHILIP
Department of Archaeology, Durham University
South Road, Durham
DH1 3LE, United Kingdom
Email: graham.philip@durham.ac.uk

REBECCA REDFERN
Centre for Human Bioarchaeology, Museum of London
150 London Wall, London
EC2Y 5HN, United Kingdom
Email: rredfern@museumoflondon.org.uk

CHARLOTTE ROBERTS
Department of Archaeology, Durham University
South Road, Durham
DH1 3LE, United Kingdom
Email: c.a.roberts@durham.ac.uk

CHRIS SCARRE
Department of Archaeology, Durham University
South Road, Durham
DH1 3LE, United Kingdom
Email: chris.scarre@durham.ac.uk

MARTIN J. SERGEANT
Division of Microbiology & Infection
University of Warwick, Coventry
CV4 7AL, United Kingdom
Email: M.J.Sergeant@warwick.ac.uk

MARK SPIGELMAN
Kuvin Center for the Study of Infectious and Tropical
Diseases and Ancient DNA
Hadassah Medical School, Hebrew University
Jerusalem, Israel
Email: spigelman@btinternet.com

IDILKO SZIKOSSY
Department of Anthropology
Hungarian Natural History Museum
Budapest, Hungary
Email: szikossy.ildiko@nhmus.hu

SONIA ZAKRZEWSKI
Department of Archaeology, University of Southampton
Highfield, Southampton
SO17 1BF
Email: srz@soton.ac.uk

Introduction

Malin Holst and Michelle Alexander

The articles included in this volume were all presented at the 15th annual British Association for Biological Anthropology and Osteoarchaeology (BABAO) conference held at the University of York on the 13th and 15th of September 2013. Sessions were structured around three main themes, with key notes from Jim Walvin (University of York, History), Jelena Bekvalac (Centre for Human Bioarchaeology, Museum of London), Ian Barnes (Natural History Museum, London) and an open session, keynote Paul O'Higgins (Centre for Anatomical and Human Sciences, Hull York Medical School). It was the aim of the conference organisers to reflect the breadth of the 500 BABAO members and their interests in the 2013 conference themes, combining the application of a variety of analysis techniques for each theme in preference of focusing on a method-based approach. We received a considerable number of high quality abstracts that made for a dynamic conference, which was reflected in the largest number of conference delegates to date (220).

Papers presented at each of the conference sessions are represented in this volume, which contains ten papers on a range of topics. Papers by Zakrzewski and Fortes-Lima and Dugoujon were incorporated into the theme of 'Constructing Identities: Ethnicity and Migration' and both have methodological aspects. Zakrzewski takes a theoretical stance to explore the multiple identities of the body, while Fortes-Lima and Dugoujon review the multidisciplinary approaches to investigate the African origin of African American communities in French Guiana, Colombia and Brazil. The theme 'Treatment of the Body: Understanding and Portrayals' was aimed at exploring different approaches to mortuary treatment through time. Within this theme, Jay *et al.* focus on the visibility of prehistoric burial practice in Britain and the Levant (the 'invisible dead'), while Dawson's paper presents evidence for diversity in late medieval Christian burial practice in Taunton, Somerset. Three papers are incorporated in the theme 'Investigating Lifeways: Diets, Disease and Occupations'. Donoghue *et al.* focus on ancient DNA to investigate *Mycobacterium tuberculosis* from 18th century mummies from Hungary. Redfern offers a bioarchaeological perspective on military communities in Roman London and Henderson takes a methodological approach to test a faster method for recording past activity-patterns in skeletal remains. The final three papers of the volume have both archaeological and methodological aspects. McIntyre and Chamberlain have a Roman focus but look to the north, using osteological and archaeological evidence to investigate health in Roman York. O'Meara explores ostoarchaeological sampling strategies, presenting data from a large-scale sieving programme of a 19th century crypt and Burrell *et al.* detail a methodological study of estimating age of non-adults.

All the articles have been subject to a rigorous peer review process and we thank the reviewers for their time and informed comments. Those articles that we have accepted represent the highest quality research. We would also like to acknowledge the organising committee, Sophy Charlton, Emily Hellewell, Keri Rowsell with aid from Matthew Collins and Oliver Craig and the many volunteers who helped us run an enjoyable conference.

Commemoration

We would like to dedicate this volume to Don Brothwell (1933–2016), Professor and later Emeritus Professor of Human Palaeoecology at the University of York, where the 2013 BABAO conference was held. Don was a most inspiring pioneer of bioarchaeology and will be greatly missed.

Biographical Notes

Malin studied archaeology at Bournemouth (HND) and Leicester University (BA) and undertook an MSc in Osteology, Palaeopathology and Funerary Archaeology at Sheffield and Bradford Universities. She has excavated in the UK since 1987 and worked as a field archaeologist until founding her company, York Osteoarchaeology Ltd, in 2003, which specialises in the excavation and analysis of human remains. Malin started teaching at the University of York in the same year and is a lecturer in the Department of Archaeology.

Malin Holst
York Osteoarchaeology,
75 Main Street, Bishop Wilton, York, YO42 1SR.
Email: malin.holst@york.ac.uk

Michelle completed her undergraduate degree in Archaeology (BSc) at Durham University. She then went on to study an MSc Biomolecular Archaeology at Manchester and Sheffield Universities. Michelle returned to Durham for her PhD exploring medieval diet using stable isotope analysis. After a brief Postdoctoral Fellowship in ancient DNA at the University of Aberdeen including visiting fellowships at Durham and Cornell University (USA), Michelle took up the position of Lecturer in Bioarchaeology at the University of York in 2012 and focuses her research on stable isotope analysis, exploring the diet of Medieval and Post-Medieval populations in the Mediterranean and the UK.

Michelle Alexander
Department of Archaeology, University of York,
King's Manor, York, YO1 7EP

Editorial Note

The contributions published in the second volume of *Trends in Biological Anthropology* were originally submitted in 2014 and therefore do not incorporate subsequent developments in human bioarchaeological research.

Tina Jakob
Department of Archaeology,
Durham University

Matryoshki, Masks and Identities: Bioarchaeology and the Body

Sonia Zakrzewski

Bioarchaeology focusses upon the interaction between the skeletal body and the archaeology of the living and dead population. The study of bioarchaeology may therefore focus on the study of a series of identities expressed both on and by the body itself. In this sense, bioarchaeology views the body as a dynamic entity with multiple levels of interpretation and engagement. This paper considers these different identities as a series of matryoshki dolls and explores these ideas with a series of exemplar case studies.

Keywords Identity; Ethnicity; Demography; Disability; Body

1. Introduction

Bioarchaeology is commonly considered to comprise the study of human remains within a broader archaeological context. This paper considers the possibilities of studying aspects of identity from the human body as expressed through biological mechanisms. Despite some notable exceptions, many bioarchaeological studies have primarily been single-issue studies of identity (Meskell, 2001), and have usually focused on palaeopathology. Similarly to Gowland and Thompson (2013), I argue here that bioarchaeology should instead consider how skeletal or mummified evidence for disease, trauma, or other aspects of identity might be evaluated as part of a synthetic holistic bodily entity, and, from this, the potential impact on both the individual person and on his/her peers further explored. Gamble (2007) has described Hawkes' (1954) ladder of inference as an 'onion of inference'. Following this analogy, I argue here that bioarchaeological identity comprises a series of overlapping existences which can be viewed as matryoshki dolls. Each matryoshka doll comprises an aspect of identity which might be opened to explore and disentangle the biological inferences underneath.

2. Bioarchaeology and Identity: What is Identity?

The body is central in the forging of identities, and study of this multiply layered body, with its own materiality, enables identities to be disentangled from the archaeological record. Bioarchaeology and osteoarchaeology study the people rather than just the palaeopathological lesions. In this sense, bioarchaeology uniquely comprises the study of people living in the past, using archaeological approaches, but situated within a broader framework developed from biological methods. In this sense, in contrast to Marty Rubin, the blogger who argued that each person has a story[1], for bioarchaeologists, each individual has multiple stories. These multiple stories, reflecting the embodied individual, comprise multiple matryoshka-like identities for that person. These many stories, however, written in varying degrees of detail, not only exist layered on top of one another, but also crosscut each other.

The archaeological or osteological body has long been seen as a source of information from which to reconstruct past human lifeways. By contrast, the 'explicit theorisation of these [skeletal or mummified] remains as the physiological embodiment of social processes and integration with social theory' (Gowland and Thompson, 2013: 3) has only developed more recently. Gowland and Thompson (2013: 2) explore the concept of the body as being more than a 'passive clothes horse for material culture'. This paper develops this approach further using examples from everyday life, albeit with a focus on Egyptian archaeology, to enable this multivocality of identity to be demonstrated. Common foci of archaeological studies of identity include gender

(Walde and Willows, 1991; Díaz-Andreu, 2005; Sofaer, 2006; Knüsel, 2011), age (Moore and Scott, 1997; Lucy, 2005a; Sofaer, 2006; Prowse *et al.*, 2007), personhood (Fowler, 2004; Boutin, 2011), power, rank or social status (Wason, 1994; Babić, 2005; Knüsel, 2011; Stodder, 2011), religion (Scott, 2011), sexuality (Dowson, 2000, 2008; Schmidt and Voss, 2000), ethnicity (Jones, 1997; Tyson Smith, 2003; Lucy, 2005b; Lozada, 2011; Zakrzewski, 2011), diet and culinary practice (Ambrose *et al.*, 2003; Eriksson *et al.*, 2008) and/or disability (Hubert, 2000a; 2000b). Despite some important and notable exceptions, such as specific papers in Powell *et al.*, (1991), Grauer and Stuart-Macadam (1998), Steckel and Rose (2002), Gowland and Knüsel (2006), Knudson and Stojanowski (2009) and Baadsgaard *et al.* (2011), these multiple strands of identity have rarely been integrated within bioarchaeology, although the importance of such an approach has been highlighted, most notably by Gowland and Thompson (2013). Bioarchaeology, palaeopathology and funerary archaeology can act as mechanisms by which these different categories can be coalesced to form overarching but multiple identities.

3.   Egypt and Bioarchaeological Identity

The majority of this paper will employ examples deriving from the Ptolemaic-Roman delta cemetery of Quesna in order to exemplify aspects of identities. The site of Quesna lies in Minufiyeh province to the north of Cairo within the Egyptian delta. The necropolis is primarily Ptolemaic and Roman in date (*c*. 332 BC–AD 641[2]), with most individuals interred in simple pit graves dug straight into the sand (for details, see Rowland, 2008; Rowland *et al.*, 2010). The actual orientation of the burials varies; most individuals are oriented east–west with their heads positioned towards the west, with a smaller proportion buried north–south, primarily with their heads to the north. Although most of the burials are of single inhumations, there are also people interred within multiple burials – sometimes with several individuals buried in one depositional event, whereas others are buried in separated and discrete burials within the same tomb. A relatively small proportion of the skeletal assemblage derives from burial in ceramic coffins. This diversity and patterning of mortuary treatment and variation in the associated funerary architecture hints at differing aspects of identity being recognised and acted upon by the burying (i.e. surviving and living) population.

3.1.   Demographic Identities

At Quesna, a life course approach has been employed to understand demographic identity. Rather than simply count the number of individuals within specific age categories, such as young adult or adolescent, the approach has been to develop a holistic series of individual osteobiographies. This approach aims to develop a nuanced approach to social aspects of age, rather than focussing entirely on the accuracy and/or precision of biological age estimation.

Of the burials excavated at Quesna, only just over half the adults excavated could be assigned into even broad age categories, such as young adult, middle-aged adult or older adult. Many were simply classified as being 'adult'. These individuals, however, each had their own distinct life and lived experiences, some of which may be reflected osteologically. For example, did any of these people experience pain during their lives, such as from arthritis in their joints? Can their biological age, currently noted simply as 'adult', be linked to and integrated with their funerary context so as to gauge their relative social age within the assemblage?

One burial demonstrates the validity of such a life-course approach to age and osteobiography. Skeleton B6, buried in an anthropoid mudbrick grave, was the best preserved of all the inhumations. The skeleton is that of an edentulous old female (Rowland 2008). She was relatively short (only approximately 142 cm tall) and exhibited an enlarged nutrient foramen in her right first metatarsal. She had a septal aperture and exhibited osteophytes on many of her vertebrae. The very fact that she survived so long, despite her lack of teeth and apparent osteoarthritis, suggests that she was 'cared for', and was thus considered to be a valued and potentially important member of the community. She was thus a 'person', possibly viewed as an 'older woman' or 'old lady', rather than simply 'being' an individual. One might also argue that she 'lived' rather than simply 'survived'.

Children have started to become relatively well contextualised within bioarchaeology (Sofaer Derevenski, 1997; Lewis, 2007, 2011). In Egyptian archaeology, despite some notable exceptions (Power, 2012; Wheeler, 2012; Wheeler *et al.*, 2013), the bioarchaeology of children and the link with children's multiple identities is still not much explored. At what biological, physiological or skeletal age are children socialised into being 'people'? Is there a recognised social period of childhood, and can this be identified from the mortuary bioarchaeology? Such an approach is being attempted at Quesna, although at present the juvenile skeletal sample is still too small to develop these ideas greatly. There are, however, two juvenile inhumations, burials B21 and B26, which are of particular note in terms of their identities. These two juveniles have very different funerary contexts, but share aspects marking their identities as being of particular importance to the local community.

B26 was found lying supine in the top layer of a mud-brick multiple burial (Rowland, 2008). Five other inhumations were also recovered from the burial structure, with each layer of burials having all the inhumations facing the same but alternate directions. B26 was a juvenile, with almost all epiphyses unfused. The only fused bones were the neural arches of the vertebrae and the innominates; the sacral bodies, heads of ribs, humerus heads, distal radial epiphyses, spinous processes and endplates of the vertebrae were all unfused. B26 lacked a skull, potentially as a result of grave robbing. The individual was estimated to have been 15–24 years old at death on the basis of pubic symphysis morphology following the Suchey-Brooks method (Brooks and Suchey, 1990). As noted above, the head was missing, and hence dental wear could not be used to aid in age estimation. B26, however, possessed many epiphyses, which, by age 15–24, should have either fused or have started to fuse together. Furthermore, the long bones were extremely long. Given the length and relative robustness of these long bones, the individual was assumed to be male. This sex assignation remains uncertain as this was the skeleton of a juvenile, but, if male, his height was calculated to have been approximately 1.7 m (following Raxter *et al.*, 2008).

Unlike B26, B21 was a single burial in a simple pit grave, oriented north–south, and cut directly into the sand (Rowland, 2008). Like B26, the body was missing most of the skull, most likely as a result of past grave robbing activity, and like B26, this inhumation had completely unfused epiphyses. Based on the pubic symphysis, age was again estimated as 15–24 years (Brooks and Suchey, 1990). The dental wear and development was also found to be similar (Brothwell, 1981; AlQahtani *et al.*, 2010). B21, like B26, had many epiphyses, such as the heads of the metacarpals, metatarsals and the proximal phalanges, that should, by 15–24 years, have either fused or have been in the process of fusing. The long bones were also relatively long, providing a height estimate of almost 1.6 m if female and almost 1.65 m if male (following Raxter *et al.*, 2008).

Both these juvenile burials were found with grave goods, although those from B26, a Ptolemaic pot sherd and a sherd incised with a *wedjat* eye (Rowland, 2008), were found in the grave fill rather than directly associated with the body. B21 was found with a Hathor plaque, a winged bird collar, a large scarab, several small scarabs, a Djed pillar and a diverse array of other amulets, plaques and pieces of cartonnage (Rowland, 2008). Both these burials were of unusually tall subadults who were still growing at time of death. This delay in epiphyseal fusion and continued growth might result from disruptions or changes to endocrine hormone levels within the body, such as somatotropin (Ortner and Putschar, 1981; Aufderheide and Rodríguez-Martín, 1998). The actual cause, such as castration, is not of importance, but rather it is the very fact that these two people were afforded non-normal burial which is. This implies that both juveniles were considered important people within the community. They may have been recognised as 'different' in some way, but, if so, this was simply one part of their multiple identities.

3.2. Race and Ethnicity

In the popular media, ethnicity and population affiliation or affinity are frequently conflated with aspects of skin colouration to form reified static race constructs. These widespread common-or-garden 'race' constructs develop from late 19th- and early 20th-century anthropological thought, which classified individuals into anatomical shape complexes and viewed these as fixed and discrete entities. This thought construct itself derives partially from

a basic understanding of Darwinism and evolutionary processes, and partially from the use of skin colour as one of the main characteristics used for ethnic differentiation as early as the 17th century (Benthien, 2002). Developing this approach, anatomical and morphological patterning has been directly mapped, in such folk taxonomies of race, onto skin colour as a layer of identity. Despite the continuous range of colouring in skin pigmentation (Relethford, 2009), these discrete folk taxonomy entities are usually viewed in binary distinctions, and so are simplified into 'Black' and 'White' (Shanklin, 1998). For a concise discussion of some of the issues surrounding skin and race, see Gowland and Thompson (2013).

By contrast, academically, ethnicity is usually viewed as being dynamic and fluid (Jones, 1997; Lucy, 2005b; Zakrzewski, 2011). It is relational and recognition depends upon personal and individual viewpoints. This is in direct contrast to early anthropological studies using 'racial types', e.g. Morton (1839, 1844) or Nott and Gliddon (1854). More recent bioarchaeological studies (Buzon, 2006, 2008; Lozada, 2011; Zakrzewski, 2011) have approached ethnicity by integrating aspects of funerary archaeology, mortuary or other cultural patterning with the biological anthropology and population affinity, and have viewed ethnicity as requiring active participation and performance. In this sense, ethnicity is one part of the social process, but may, at times, be underpinned by some degree of biological arrangement. This approach follows Jones (1997: 84) in arguing that 'ethnic groups are culturally ascribed identity groups, which are based on the expression of a real or assumed shared culture and common descent'. Ethnicity is thus malleable but also requires an implicit consciousness of difference. Bioarchaeologically, this usually means that ethnicity is derived from patterning in grave goods, funerary architecture and burial practice, such as at Tombos in Sudan (Buzon, 2006, 2008) or in the Nile Valley at Gebelein (Zakrzewski, 2001). In this latter case, aspects of the individual identification within a self-defining system (Jones, 1997) is of particular note as, at this site, Nubian mercenaries who had married Egyptian women were buried in an Egyptian manner, but were depicted as *nHsy*, the ancient Egyptian name for Nubians, thereby retaining their ethnic identity. Unfortunately at the delta site of Quesna, the broader archaeology is too poorly preserved in association with the burials for aspects of ethnicity to be modelled, but one might expect that differentiation in funerary form, such as the differences between people buried in ceramic coffins relative to simple pit graves, or in burial orientation, might link to aspects of ethnicity or layer of identity.

3.3. Body Size and Shape

Height and body shape have long been studied within bioarchaeology, having been used widely as indicators of socioeconomic status and well-being (Bogin 2001), and both clearly exist as layers of identity. We simply need to see current western magazine covers to be aware of the importance of such dynamics within identity and the modern fascination with size zero body shapes etc. But if we are studying past concepts of identity, it is more important to recognise the changes and variation in importance of body size and shape in past groups. In this sense, one can compare these modern western magazine covers with their idealised view of the body beautiful with Rubenesque ideals.

Ancient Egyptians recognised differences in body size and shape between individuals and groups, and there were three distinct Egyptian words for abnormally short people, including different types of dwarfs (*dng*, *nmw*, and *Hwa*) (Dasen, 1993). This implies that difference was recognised, but should not be considered as negative. Indeed the modern association of dwarfing with disability might not be appropriate for ancient Egyptian mental concepts.

At our exemplar site of Quesna, distinct differences in computed body height are seen between individuals (calculated following the methods of Raxter *et al.*, 2008). Significant sexual dimorphism in height was found, with males being almost 10% taller than females. This is greater sexual dimorphism than in other ancient Egyptian samples (Zakrzewski 2003, 2007). Stature and body size, and their associated sexual dimorphism, are linked not only to childhood health and socioeconomic status (Eveleth and Tanner, 1990; Steckel, 1995, 2009; Bogin, 1999, 2001), but are also linked to gender relations and hierarchical social organisation (Alexander *et al.*, 1979; Gray and Wolfe, 1980; Gaulin and Boster, 1985, 1992; Holden and Mace, 1999; Guégan *et al.*, 2000; Kanazawa and Novak, 2005). This implies that the high level of sexual dimorphism at Quesna may have underlying causes that could affect the local social structure and organisation. For example, the Quesna Ptolemaic-Roman population may have treated boys and girls differently, and/or polygynous marriages or other relations might have occurred.

Furthermore, the two individuals described earlier, B21 and B26, were both unusually tall (1.60–1.65 m and 1.7 m respectively) despite being juveniles who had not yet attained their skeletally mature heights. These were two of the tallest people recovered from the site and it is possible that the local community recognised that these two were juveniles who were still growing, and this aspect of their identity might have been recognised as unusual or different by the surviving population, i.e. those people actually burying them. This implication links to emic constructions of 'otherness' (Hubert, 2000b) and permits discussion of the concept of otherness in past human groupings.

4. Otherness and Concepts of Dis/Ability

What is 'other'? What is 'different'? Understanding identity in the manner articulated here requires the recognition and understanding of what constitutes embodied difference. Within bioarchaeology, too frequently we have simply equated palaeopathology with disease and expressions

of severe disease with potential disability. This has been noted and cautioned by others (see Craig and Craig, 2013 or Young and Lemaire, 2012 for recent good examples), and, as noted earlier, recognition of different need not be negative. Biological differences are relatively well represented in Egyptian art, with both representations of different 'ethnic' groups (e.g. Buzon, 2006; Nibbi, 1986) and impressions of health or disease recognisable (e.g. Iversen, 1975; Dasen, 1993; Gordon and Schwabe, 2004). As noted earlier, it is clear that individuals of abnormally short stature were considered different from 'normal' people (for detail see Weeks, 1970; Iversen, 1975; Dasen, 1993; Robins, 1994), and some of dwarfs were of high social ranking. For example, the dwarf Perniankhu is depicted with symbols of authority (Hawass, 1991; Wilkinson, 2007), and hence his dwarfing may have been seen as some form of divine marking (Dasen, 1993). This implies that dwarves, although viewed as 'other', were potentially considered able-bodied and positive members of the community (Sullivan, 2001). Depictions of such 'others' in ancient Egypt include potential representations of Pott's disease (Reeves, 1992; Filer, 1995; Halioua and Ziskind, 2005; Ziskind and Halioua, 2007) and either *talipes equinus* or poliomyelitis from New Kingdom tombs (Filer, 1995; Nunn, 1996; Halioua and Ziskind, 2005). Such physical deformities were not considered to be a major part of the individual's own persona and thus were not commonly depicted (Dasen, 1993). In addition, in certain circumstances, depictions of what might be considered disabilities, such as blindness, were employed to indicate that either the individual was of particular importance or to depict their piety (Dasen, 1993). One such individual, a singer from the Ramesside period, was depicted as blind when playing music for his patron deities, but sighted or at least with his eyes open in other images (Dasen, 1993). This suggests an understanding of the multiplicity of identities and a duality to his individual social persona.

These diverse forms of disability demonstrate that our interpretation of disability is somewhat limited and may instead rather reveal our own personal visceral responses to perceived 'otherness'. Disability itself is thus difficult to define. The World Health Organisation defines disability as 'an umbrella term, covering impairments, activity limitations, and participation restrictions' (WHO, nd). As a result, disability or disabilities may appear sporadically, and sometimes chronically, throughout a person's lifetime. Bioarchaeology has only recently started to adopt the diverse disabilities literatures, and hence these approaches are only now starting to be integrated into concepts of past life courses. If disability is viewed in terms of the social restrictions placed upon individuals with bodily impairment, following Oliver (1983), disability is simply a form of limited activity. Taking this approach, a disabled person is any person who has some condition that inhibits them undertaking the full range of activities considered normal at their own age (Thomas, 2007). This focuses the importance on the actual 'living with the condition', rather than the condition in itself. Furthermore, this implies that disability can be age-related and might be viewed as rather more universally experienced in terms of degrees of expressions of ability. There is therefore a fluid boundary between disabled and able-bodied, with each shading and gradations in the self-identification along the continuum of disability depending on the form of activity being undertaken or attempted. Concepts of disability or ability thus each can be viewed as a different matryoshka of identity for each person.

Developing this approach, and especially in association with the 'restriction on participation' from the WHO definition, disability is a complex series of phenomena that reflect the interaction between the person and the local society (Zakrzewski, 2014). I have argued elsewhere (Zakrzewski, 2015) that at certain times in their life course, a person may experience temporary activity restrictions. Examples include temporary reductions in mobility resulting from long bone fracture and from symphysis pubis dysfunction during pregnancy. However, in addition to these likely relatively visible impairments, other less visible impingements might also exist. Examples could include blindness, such as argued for Shanidar I (Trinkaus and Zimmerman, 1982) or deafness, as for the congenitally deaf child from Poundbury (Farwell and Molleson, 1993). We must, however, consider how such 'disabilities' or rather reduced abilities impinge on the individuals themselves and their close community. Within my own family, I am well-known for being almost blind without my spectacles. But it is not just my vision that is impaired. Without my spectacles, I am 'unable' to hear and interpret information. Thus my personal reliance on vision affects my abilities to function using other sensory pathways.

What about the people at Quesna? Did any of the individuals buried in the cemetery have reduced abilities? Certainly many of them exhibited osteophytosis and degenerative joint disease. Furthermore, osteoarthritic lesions on bones of the hands and feet were relatively commonly found, and affected people might have found that some activities were difficult to undertake, thereby reducing their participation. Scoliosis was also noted in some Quesna individuals, including B11 (that of an adult male). Scoliosis may affect the heart and reduce lung capacity and, in severe cases, lead to cardiac or respiratory failure (Nilsonne and Lundgren, 1968; Pehrsson *et al.*, 1991), thus affected individuals can easily be envisaged as being less able to undertake physical work, and thus be relatively disabled in comparison with contemporaries. Rather than simply considering the palaeopathology of each individual, as bioarchaeologists, I would argue, we need to think of people living along some continuum of varying ability, and it is this dis/ability that impacts on the rest of the local society. People may move forwards and backwards along that dis/ability continuum at different life stages, but this is simply one facet or matryoshka of their identity. This is of particular importance in certain past groups, as some, such as ancient Egypt, appear relatively accepting of individuals considered (as)

'different' or 'other' (Jeffreys and Tait, 2000); indeed 'What is perceived as a "disability" or as "madness" in one society, in another may be considered as just one attribute among many which make up an individual, or may not be perceived as part of the individual at all' (Waldron, 2000: 7).

5. Bioarchaeology, Matryoshki and Masks of Identity

I have attempted here to demonstrate how bioarchaeology can be so much more than simply a catalogue of burials, of individuals and their palaeopathologies. Gamble (2007: 89–90) has argued that Hawkes' (1954) ladder of inference is better considered as an 'onion of inference'. Given that in bioarchaeology we are studying the individual people themselves, I believe that an analogy with matryoshki dolls is more appropriate in that each person comprises multiple layers of identity that overlap each other. Some of these layers are more hidden or more masked at certain periods of life, whereas others are very visible. Certain of these can be clearly envisaged as overlying and larger dolls, such as the determined wearing of clothing, whereas others are more internal, hidden and personal. In my own personal experience, the former might include the determined wearing of my Newcastle United football shirt, whereas the latter more internal or personal aspects might include aspects of my identity as a parent. Personal concepts of ethnicity or dis/ability might form distinct matryoshki dolls, with their actual positioning relative to the outermost doll varying depending on factors including the geographic, temporal or sociocultural situation. Each matryoshka doll thus provides its own series of identity inferences.

As bioarchaeologists, we are all aware of the potential of human remains, but we need to break out of the burial catalogue in the understandings of others and form part of the social critique in archaeology. Bioarchaeology, especially when integrated with archaeothanatology (following Duday, 2009) provides the chance to understand the impact upon both the person themselves and their peers of their disease, trauma, dis/ability, population affinity and ethnicity etc. One mechanism by which to do this might be to develop multiple osteobiographies, and use these in association with more traditional population- or sample-based approaches. Each person has their own individual biography, and together these form parallel osteobiographies for the group. Each biography will incorporate diverse layers of the various individuals' identities, linking and imprinting their biological bodies with their social beings.

I have argued recently that osteobiographies can be synthesised with population- or sample-based approaches (Zakrzewski, 2015). This approach builds on the excellent model developed by Robb (2002). In his study, Robb provides one of the best demonstrations of the potential for osteobiography in developing understanding of the individual and identity. Boutin (2011) has further developed the approach by linking osteobiography with personhood. I would argue that synthesising studies of the individual (for best practice, see papers in Stodder and Palkovich, 2012) with the more traditional and statistically based analyses of palaeopathology, population affinity or archaeothanatology can lead to the uncovering of multiple identities layered one on top of another in a similar manner to each matryoshka in a set of dolls. In this sense, identity forms the embodied relationship performed between the individual and the outside world. By its very bodily nature, bioarchaeology can elucidate and tease out these multiple aspects of identity bound together in constructions of self and other.

Notes

1   'Behind every mask there is a face, and behind that a story.' Marty Rubin, 2010. http://mydailyaphorism. blogspot.co.uk/2010/05/behind-every-mask-there-is-face-and.html (accessed November, 2013).
2   The burials have been dated by seriation methods, using ceramic sherds, as export of samples for radiocarbon dating is not permitted.

References

Alexander, R.D., Hoogland, J.L., Howard, R.D., Noonan, K.M., Sherman, P.W. 1979. Sexual dimorphism and breeding systems in pinnipeds, ungulates, primates and humans, in: Chagnon, N.A., Irons, W. (Eds.), *Evolutionary Biology and Human Behaviour: An Anthropological Perspective.* Duxbury Press, North Scituate, MA, pp. 402–435.

AlQahtani, S.J., Hector, M.P., Liversidge, H.M. 2010. Brief communication: The London atlas of human tooth development and eruption. *American Journal of Physical Anthropology* 142, 481–490.

Ambrose, S., Buikstra, J., Harold, W. 2003. Status and gender differences in diet at Mound 72, Cahokia, revealed by isotopic analysis of bone. *Journal of Archaeological Science* 22, 217–226.

Aufderheide, A.C., Rodríguez-Martín, C. 1998. *The Cambridge Encyclopedia of Human Paleopathology.* Cambridge University Press, Cambridge.

Baadsgaard, A., Boutin A.T., Buikstra, J.E. (Eds.), 2011. *Breathing New Life into the Evidence of Death: Contemporary Approaches to Bioarchaeology.* School for Advanced Research Press, Santa Fe.

Babić, S. 2005. Status identity and archaeology, in: Díaz-Andreu, M., Lucy, S., Babić, S., Edwards, D.N. (Eds.), *The Archaeology of Identity.* Routledge, London, pp. 67–85.

Benthien, C. 2002. *Skin: On the Cultural Border Between Self and World: On the Cultural Border Between Self and the World.* Columbia University Press, New York.

Bogin, B. 1999. *Patterns of Human Growth*, 2nd edition. Cambridge University Press, Cambridge.

Bogin, B. 2001. *The Growth of Humanity.* Wiley-Liss, New York.

Boutin, A.T. 2011. Crafting a bioarchaeology of personhood: Osteobiographical narratives from Alalakh, in: Baadsgaard, A., Boutin A.T., Buikstra, J.E. (Eds.), *Breathing New Life into the Evidence of Death: Contemporary Approaches to Bioarchaeology.* School for Advanced Research Press, Santa Fe, pp. 109–133.

Brooks, S., Suchey, J.M. 1990. Skeletal age determination based on the os pubis: A comparison of the Ascádi-Nemeskéri and Suchey–Brooks methods. *Human Evolution* 5, 227–238.

Brothwell, D. 1981. *Digging Up Bones.* Cornell University Press, Ithaca.

Buzon, M.R. 2006. The relationship between biological and ethnic identity in New Kingdom Nubia. *Current Anthropology* 47, 683–695.

Buzon, M.R. 2008. A bioarchaeological perspective on Egyptian colonialism in Nubia during the New Kingdom. *Journal of Egyptian Archaeology* 94, 165–181.

Craig, E., Craig, G. 2013. The diagnosis and context of a facial deformity from an Anglo-Saxon Cemetery at Spofforth, North Yorkshire. *International Journal of Osteoarchaeology* 23, 631–639.

Dasen, V. 1993. *Dwarfs in Ancient Egypt and Greece, Oxford Monographs in Classical Archaeology.* Clarendon Press, Oxford.

Díaz-Andreu, M. 2005. Gender identity, in: Díaz-Andreu, M., Lucy, S., Babić, S., Edwards, D.N. (Eds.), *The Archaeology of Identity.* Routledge, London, pp. 13–42.

Dowson, T. (Ed.), 2000. Queer archaeologies. *World Archaeology* 32(2). Special Issue.

Dowson, T. 2008. Queering sex and gender in Ancient Egypt, in: Graves-Brown, C. (Ed.), *Sex and Gender in Ancient Egypt. "Don your wig for a joyful hour".* The Classical Press of Wales, Swansea. pp. 27–46.

Duday, H. 2009. *The Archaeology of the Dead: Lectures in Archaeothanatology*, trans. A.M. Cipriani and J. Pearce. Oxbow Books, Oxford.

Eveleth, P.B., Tanner, J.M. 1990. *Worldwide Variation in Human Growth*, 2nd edition. Cambridge University Press, Cambridge.

Eriksson, G., Linderholm, A., Fornander, E., Kanstrup, M., Schoultz, P., Olofsson, H., Lidén, K. 2008. Same island, different diet: Cultural evolution of food practice on Öland, Sweden, from the Mesolithic to the Roman Period. *Journal of Anthropological Archaeology* 27, 520–543.

Farwell, D.E., Molleson, T.I. 1993. *Excavations at Poundbury 1966–80, Volume II: The Cemeteries.* Dorset Natural History and Archaeological Society, Dorset.

Filer, J. 1995. *Egyptian Bookshelf: Disease.* British Museum Press, London.

Fowler, C. 2004. *The Archaeology of Personhood: An Anthropological Approach.* Routledge, London.

Gamble, C. 2007. *Archaeology: The Basics*, 2nd edition. Routledge, Abingdon.

Gaulin, S., Boster, J. 1985. Cross-cultural differences in sexual dimorphism: is there any variance to be explained? *Ethology and Sociobiology* 6, 219–225.

Gaulin, S.J.C., Boster, J.S. 1992. Human marriage systems and sexual dimorphism in stature. *American Journal of Physical Anthropology* 89, 467–475.

Gordon A.H., Schwabe, C.W. 2004. *The Quick and the Dead: Biomedical Theory in Ancient Egypt.* Brill Styx, Leiden.

Gowland, R. 2006. Age as an aspect of social identity: The archaeological funerary evidence, in: Gowland, R., Knüsel, C. (Eds.), *Social Archaeology of Funerary Remains.* Oxbow Books, Oxford, pp. 143–154.

Gowland, R., Knüsel, C. (Eds.), 2006. *Social Archaeology of Funerary Remains.* Oxbow Books, Oxford.

Gowland, R.L., Thompson, T.J.U. 2013. *Human Identity and Identification.* Cambridge University Press, Cambridge.

Grauer, A.L., Stuart-Macadam, P. (Eds.), 1998. *Sex and Gender in Paleopathological Perspective.* Cambridge University Press, Cambridge.

Gray, J.P., Wolfe, L.D. 1980. Height and sexual dimorphism of stature among human societies. *American Journal of Physical Anthropology* 53, 441–456.

Guégan, J-F., Teriokhin, A.T., Thomas, F. 2000. Human fertility variation, size-related obstetrical performance and the evolution of sexual stature dimorphism. *Proceedings of the Royal Society London B, Biological Sciences* 267, 2529–2535.

Halioua, B., Ziskind, B. 2005. *Medicine in the Days of the Pharaohs,* trans. M.B. DeBevoise. Harvard University Press, Cambridge.

Hawass, Z. 1991. The statue of the Dwarf Pr-n(j)-'nḫ(w) recently discovered at Giza. *Mitteilungen des Deutschen Arch*äologischen *Instituts Abteilung Kairo* 47, 157–162.

Hawkes, C. 1954. Archaeological theory and method: Some suggestions from the old world. *American Anthropology* 56, 155–168.

Holden, C., Mace, R. 1999. Sexual dimorphism in stature and women's work: A phylogenetic cross-cultural comparison. *American Journal of Physical Anthropology* 110, 27–45.

Hubert, J. (Ed.), 2000a. *Madness, Disability and Social Exclusion,* One World Archaeology 40. Routledge, London.

Hubert, J. 2000b. Introduction: the complexity of boundedness and exclusion, in: Hubert, J. (Ed.) *Madness, Disability and Social Exclusion,* One World Archaeology 40. Routledge, London, pp. 1–8.

Iversen, E. 1975. *Canon and Proportions in Egyptian Art.* Aris & Phillips, Warminster.

Jeffreys, D., Tait, J. 2000. Disability, madness, and social exclusion in Dynastic Egypt, in: Hubert, J. (Ed.), *Madness, Disability and Social Exclusion,* One World Archaeology 40. Routledge, London, pp. 87–95.

Jones, S. 1997. *The Archaeology of Ethnicity.* Routledge, London.

Kanazawa, S., Novak, D.L. 2005. Human sexual dimorphism in size may be triggered by environmental cues. *Journal of Biosocial Science* 37, 657–665.

Knudson, K.J., Stojanowski, C.M. (Eds.), 2009. *Bioarchaeology and Identity in the Americas.* University Press of Florida, Gainesville.

Knüsel, C.J. 2011. Men take up arms for war: Sex and status distinctions of humeral medial epicondylar avulsion fractures in the archaeological record, in: Baadsgaard, A., Boutin A.T., Buikstra, J.E. (Eds.)*, Breathing New Life into the Evidence of Death: Contemporary Approaches to Bioarchaeology.* School for Advanced Research Press, Santa Fe, pp. 221–249.

Lewis, M.E. 2007. *The Bioarchaeology of Children.* Cambridge University Press, Cambridge.

Lewis, M. 2011. The osteology of infancy and childhood: Misconceptions and potential, in: Lally, M., Moore, A. (Eds.), *(Re)thinking the Little Ancestor: New Perspectives on the Archaeology of Infancy and Childhood.* BAR International Series (S2271). Archaeopress, Oxford, pp. 1–13.

Lozada, M.C. 2011. Cultural determinants of ancestry: A lesson for studies of biological relatedness and ethnicity in the past, in: Baadsgaard, A., Boutin A.T., Buikstra, J.E. (Eds.), *Breathing New Life into the Evidence of Death: Contemporary Approaches to Bioarchaeology.* School for Advanced Research Press, Santa Fe, pp. 135–149.

Lucy, S. 2005a. The archaeology of age, in: Díaz-Andreu, M., Lucy, S., Babić, S., Edwards, D.N. (Eds.), *The Archaeology of Identity.* Routledge, London, pp. 43–66.

Lucy, S. 2005b. Ethnic and cultural identities, in: Díaz-Andreu, M., Lucy, S., Babić, S., Edwards, D.N. (Eds.), *The Archaeology of Identity.* Routledge, London, pp. 86–109.

Moore, J., Scott, E. (Eds.), 1997. *Invisible People and Processes.* Leicester University Press, London.

Morton, S.G. 1839. *Crania Americana; or, A comparative view of the skulls of various Aboriginal Nations of North and South America: To which is prefixed an essay on the varieties of the human species.* J. Dobson, Philadelphia.

Morton, S.G. 1844. *Crania Aegyptiaca; or, Observations on Egyptian ethnography, derived from anatomy, history, and the monuments.* J. Penington, Philadelphia.

Nibbi, A. 1986. *Lapwings and Libyans in Ancient Egypt.* Discussions in Egyptology publications, Oxford.

Nilsonne, U., Lundgren, K.-D. 1968. Long-term prognosis in idiopathic scoliosis. *Acta Orthopaedica Scandinavica* 39, 456–465.

Nott, J.C., Gliddon, G.R. 1854. *Types of mankind: or, Ethnological researches, based upon the ancient monuments, paintings, sculptures, and crania of races, and upon their natural, geographical, philological and Biblical history: illustrated by selections from the inedited papers of Samuel George Morton ... and by additional contributions from Prof. L. Agassiz, W. Usher, and Prof. H. S. Patterson.* Lippincott, Gramoo & Co, Philadelphia.

Nunn, J.F. 1996. *Ancient Egyptian Medicine.* British Museum Press, London.

Oliver, M. 1983. *Social Work with Disabled People.* Macmillan, Basingstoke.

Ortner, D.J., Putschar, W.G.J. 1981. *Identification of Pathological Conditions in Human Skeletal Remains.* Smithsonian Institution Press, Washington.

Pehrsson, K., Bake, B., Larsson, S., Nachemson, A. 1991. Lung function in adult idiopathic scoliosis: A 20 year follow up. *Thorax* 46, 474–478

Powell, M.L, Bridges, P.S., Wagner Mires, A.M. (Eds.), 1991. *What Mean These Bones? Studies in Southeastern Bioarchaeology.* University of Alabama Press, Tuscaloosa.

Power, R. 2012. *From the Cradle to the Grave: Child, Infant and Foetal Burials in the Egyptian Archaeological Record from the Early Dynastic Period to the Middle Kingdom (ca. 3300-1650 BC).* Unpublished PhD Thesis, Macquarie University.

Prowse, T.L., Schwarcz, H.P., Garnsey, P., Knyf, M., Macchiarelli, R., Bondioli, L. 2007. Isotopic evidence for age-related immigration to Imperial Rome. *American Journal of Physical Anthropology* 132, 510–519.

Raxter, M., Ruff, C., Azab, A., Erfan, M., Soliman, M., El-Sawaf, A. 2008. Stature estimation in ancient Egyptians: a new technique based on anatomical reconstruction of stature. *American Journal of Physical Anthropology* 136, 147–155.

Reeves, C. 1992. *Egyptian Medicine.* Shire, Princes Risborough.

Robb, J. 2002. Time and biography: Osteobiography of the Italian Neolithic lifespan, in: Hamilakis, Y., Pluciennik, M., Tarlow, S. (Eds.), *Thinking through the Body: Archaeologies of Corporeality.* Kluwer / Plenum, New York, pp. 153–171.

Robins, G. 1994. *Proportion and Style in Ancient Egyptian Art.* University of Texas Press, Austin.

Rowland, J. 2008. The Ptolemaic-Roman cemetery at the Quesna archaeological area. *Journal of Egyptian Archaeology* 94, 69–93.

Rowland, J., Inskip, S., Zakrzewski, S.R. 2010. The Ptolemaic-Roman cemetery at the Quesna archaeological area. *Journal of Egyptian Archaeology* 96, 31–48.

Schmidt, R.A., Voss, B.L. (Eds.), 2000. *Archaeologies of Sexuality.* Routledge, London.

Scott, R.E. 2011. Religious identity and mortuary practice: The significance of Christian burials in early Medieval Ireland, in: Baadsgaard, A., Boutin A.T., Buikstra, J.E. (Eds.), *Breathing New Life into the Evidence of Death: Contemporary Approaches to Bioarchaeology.* School for Advanced Research Press, Santa Fe, pp. 55–77.

Shanklin, E. 1998. The profession of the color blind: Sociocultural anthropology and racism in the 21st Century. *American Anthropology* 100, 669–679.

Sofaer Derevenski, J. 1997. Engendering children, engendering archaeology, in: Moore, J., Scott, E. (Eds.), *Invisible People and Processes.* Leicester University Press, London, pp. 192–202.

Sofaer, J. 2006. *The Body as Material Culture.* Cambridge University Press, Cambridge.

Steckel, R.H. 1995. Stature and the standard of living. *Journal of Economic Literature* 33, 1903–1940.

Steckel, R.H. 2009. Heights and human welfare: Recent developments and new directions. *Explorations in Economic History* 46, 1–23.

Steckel, R.H., Rose, J.C. (Eds.), 2002. *The Backbone of History.* Cambridge University Press, Cambridge.

Stodder, A.L.W., Palkovich, A.M. (Eds.), 2012. *The Bioarchaeology of Individuals.* University Press of Florida, Gainesville.

Sullivan, R. 2001. Deformity: A modern western prejudice with ancient origins. *Proceedings of the Royal College of Physicians of Edinburgh* 31, 262–266.

Thomas, C. 2007. *Sociologies of Disability and Illness.* Palgrave Macmillan, Basingstoke.

Trinkaus, E., Zimmerman, M.R. 1982. Trauma among the Shanidar Neandertals. *American Journal of Physical Anthropology* 57, 61–76.

Walde, I.D., Willows, N.D. (Eds.), 1991. *The Archaeology of Gender.* The University of Calgary, Calgary.

Waldron, T. 2000. Hidden or overlooked? Where are the disadvantaged in the skeletal record?, in: Hubert, J. (Ed.), *Madness, Disability and Social Exclusion.* One World Archaeology 40. Routledge, London, pp. 29–45.

Wason, P.K. 1994. *The Archaeology of Rank.* Cambridge University Press, Cambridge.

Weeks, K.R. 1970. *The Anatomical Knowledge of the Ancient Egyptians and the Representation of the Human Figure in Egyptian Art.* Unpublished PhD Thesis, Yale University.

Wheeler, S.M. 2012. Nutritional and disease stress in juveniles from the Dakhleh Oasis, Egypt. *International Journal of Osteoarchaeology* 22, 219–234.

Wheeler, S.M., Williams, L., Beauchesne, P. Dupras, T.L. 2013. Shattered lives and broken childhoods: Evidence of physical child abuse in ancient Egypt. *International Journal of Paleopathology* 3, 71–82.

WHO, nd. *Disabilities.* http://www.who.int/topics/disabilities/en/ (accessed on 20th August, 2012)

Wilkinson, T. 2007. *Lives of the Ancient Egyptians.* Thames & Hudson, London

Zakrzewski, S.R. 2001. *Continuity and Change: A Biological History of Ancient Egypt.* Unpublished PhD Thesis, University of Cambridge.

Zakrzewski, S.R. 2003. Variation in ancient Egyptian stature and body proportions. *American Journal of Physical Anthropology* 121, 219–229.

Zakrzewski, S.R. 2007. Gender relations and social organisation in the Predynastic and early Dynastic periods, in: Goyon, J.-Cl., Cardin, C. (Eds.), *Proceedings of the 9th International Congress of Egyptologists.* Orientalia Lovaniensia Analecta 150, Peeters, Leuven, pp. 2005–2019.

Zakrzewski, S.R. 2011. Population migration, variation and identity: An Islamic population in Iberia, in: Agarwal, S.C., Glencross, B.A. (Eds.), *Social Bioarchaeology.* Blackwell Studies in Global Archaeology, Wiley-Blackwell, Chichester, pp. 183–211.

Zakrzewski, S.R. 2014. Palaeopathology, disability and bodily impairments, in: Metcalfe, R., Cockitt, J. (Eds.), *Palaeopathology of Egypt and Nubia: A Century in Review.* Archaeopress Egyptology 6, Archaeopress, Oxford, pp. 57–68.

Zakrzewski, S.R. 2015. "Behind every mask there is a face, and behind that a story." Egyptian Bioarchaeology and Ancient Identities, in: Ikram, S., Walker, R., Kaiser, J. *Bioarchaeology of Ancient Egypt.* AUC Press, Cairo, pp. 157–167.

Ziskind, B., Halioua, B. 2007. La tuberculose en ancienne Égypte. *Revue Maladies Respiratoires* 24, 1277–1283.

2

Revisiting Genetic Ancestry in African Diaspora Communities from Atlantic South America

César Fortes-Lima and Jean-Michel Dugoujon

During the period of the transatlantic slave trade (15th–19th centuries) millions of people were forced to move from Africa to the Americas, dramatically changing the human genetic landscape of the New World. Today, the descendants of enslaved Africans have sought to contextualize their genetic ancestry, by finding out more about the identity and origin of their putative ancestors.

Recently, multidisciplinary approaches to the study of African ancestry explore the biological and cultural consequences of this unprecedented movement of people in modern-day African Diaspora populations using historical resources and genetic markers with powerful geographic discriminations, such as the Y-chromosome and mitochondrial DNA (Fortes-Lima, 2016; Salas et al., 2004; Stefflova et al., 2011). An interesting case study is that of the Noir Marron *population in Suriname and French Guiana; they retain a strong African cultural heritage (Price, 1996), Creole language (Huttar, 2010) and genetic history that are deeply rooted in an African past (Brucato et al., 2010; Fortes-Lima et al., 2016; 2017).*

In this review, we explore the complexity of historical, linguistic and genetic studies on the African origin of African-descendant communities from Suriname, French Guiana, Colombia and Brazil. The aim is to complement gaps in historical records relating to the slave trade in order to gain a relevant understanding of the African admixture dynamics during the slave trade in South America.

Keywords Slave trade; *Noir Marron*; Mitochondrial DNA; Y-chromosome; Admixture; Population structure

1. Introduction

1.1. The Transatlantic Slave Trade: Historical Background

The transatlantic slave trade (15th–19th centuries) dramatically changed the demography of the New World. During this period around twelve million Africans were forcely moved to American and Caribbean colonies, making it one of the most important movements of people in world history. To highlight the extent of this movement, about 400,000 slaves arrived to work in plantations in Guiana, between Brazil and Venezuela, from 1604 to 1815, mainly for Dutch, but also for English, French and Portuguese settlers (Eltis and Richardson, 2010).

According to historical resources, enslaved Africans were transported from eight major coastal regions in Sub-Saharan Africa (Figure 2.1); 6% of slaves were from Senegambia, 3% from Sierra Leone, 3% from Windward Coast, 10% from Gold Coast, 16% from Bight of Benin, 13% from Bight of Biafra, 45% from West Central Africa, 4% from Southeast Africa, and their proportion differs among North, Central and specially South America (Eltis and Richardson, 2010). However, there is a lack of data on the African ethnicity of slaves arriving to America, because (1) the largest historical databases, like the Transatlantic Slave Trade Database (Eltis, 2009), detail the departure port for the majority of slaves, but not their home town or ethnicity, and (2) the inconsistencies and inaccuracies regarding to the demographic history and migration routes of 10.7 million enslaved Africans who survived the Middle Passage and disembarked in the Americas (Eltis and Richardson, 2010).

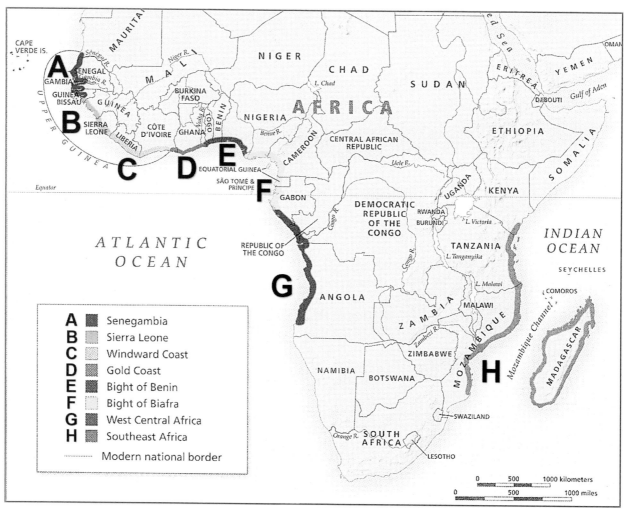

Figure 2.1 Historical African coastal regions from which slaves were carried during the slave trade. The map shows the limits of the eight major coastal regions of Sub-Saharan Africa and their alignment with modern African nations (Eltis and Richardson, 2010).

Consequently the Middle Passage was a genetic bottleneck for enslaved Africans, in that it severely restricted the African genetic diversity. Survivors of the Middle Passage were, however, able to regenerate much of this original diversity, primarily through gene flow between Africans in the New World, since African tribal and regional restrictions on marriage were not in force, and secondarily through gene flow with non-African populations. Currently, due to the significant lack of knowledge on the ancestry of African descendants, the characterisation of African genetic identity in present-day African-descendant communities in the New World are of particular scientific interest in order to better understand the impact of the slave trade.

For more than four centuries, in remote areas throughout the West Indies, Central America, South America and North America, thousands of enslaved Africans managed to escape from the plantations of European colonizers, in search of freedom. These African runaways or *Maroons* (from the Latin-American Spanish word *cimarrón*; which means wild or fugitive) formed small independent settlements (Price, 2002). These Maroon communities have subsequently emerged as free and independent societies that had forced colonial governments to sign treaties to

guarantee their freedom, their land and their political autonomy. After centuries of survival and adaptation, they have developed a unique identity and history. Today, descendants of some of the original Maroon communities still preserve a strong sense of their history, traditions, values and identity, deeply rooted in their African past. For example, *Noir Marron* populations in Suriname and French Guiana and Suriname (Figure 2.2), since their formation in the 17th century, have adapted to the Amazonian environment due to their cultural practices inherited from Africa, America and Europe (Price, 1996). A unique ethnic group has evolved, which is part of the diversity of the African Diaspora in the Americas. In most African descendants, cultural exchanges are followed by gene flow, but the *Noir Marron* have conserved a large African genetic inheritance (Fortes-Lima *et al.*, 2017).

Today, it is increasingly common for the descendants of the enslaved Africans to want to contextualize their ancestry, by understanding more about who their ancestors were, where they came from, what conditions initiated their movements within the African continent and their forced migrations to the Americas, and what New World genetic events they may have experienced (Nelson, 2016). Each of these inquiries is ultimately designed to inform

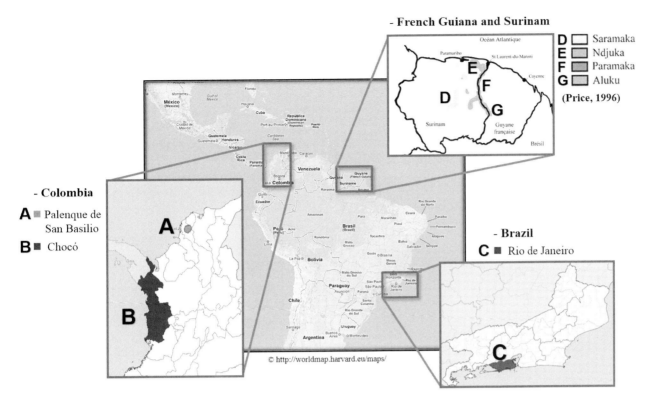

Figure 2.2 Location of studied African Diaspora communities in South America.

on the health status, ancestral background and potential disease susceptibilities of contemporary descendants (Reiner *et al.*, 2005; Jackson and Borgelin, 2010).

The historical implications of the slave trade can consequently be analysed, to create a new diasporic vision of African descendants from South America, and to address long-standing questions regarding the African origin of populations in different geographical regions. In this review we aim to show how genetic, linguistic and historical studies have identified the origin and the geographic genetic composition of current African-descendant communities in Brazil, Colombia, Suriname and French Guiana (Figure 2.2), due to their high African genetic ancestry and cultural legacy (Fortes-Lima, 2016).

1.2. Linguistic Influences in the Formation of the Creoles of Suriname and French Guiana

Linguistic studies have identified a West African origin to the mixed Creole languages currently spoken in South America (e.g., in Suriname and French Guiana), with some Gbe varieties from West Africa, including Fongbe (Lefebvre, 2013), and some other languages from the Gold Coast and Bight of Benin.

These English-based Creole languages: Ndjuka, Saramaka, Paramaka and Aluku (language names are the same as the *Noir Marron* tribe names), include lexical items from many distinct, identifiable African and European sources (Huttar, 2009, 2012). Among the African languages, slave trade records indicate the numerical dominance of Bantu speakers (e.g., Kikongo from West Central Africa) and Gbe speakers in the late 17th century,

Gbe speakers in the early 18th century, Akan speakers in the 1720s and 1730s, and 'Upper Guinea' (and to a lesser extent Bantu) speakers thereafter (Huttar, 2010). Indeed, Smith (2002) showed that during 1675–1714, the formative period of these languages, 50% of the enslaved Africans in the Guyana regions originated from the Gold Coast and the Bight of Benin. Therefore, Gbe varieties, rather than Bantu or other linguistic families, were the base language in the formation and history of these Creoles languages from Suriname and French Guiana (Huttar, 2010).

The Aluku, Paramaka, Ndjuka and Saramaka represent the four major *Noir Marron* communities in Suriname and French Guiana (Figure 2.2) and are original examples of African-descendant populations in Latin America. Despite a long history in the Amazonian rainforest, involving cultural exchanges with neighbouring Europeans and Amerindians, the marital practices probably prevented gene flow between their communities. According to a recent genetic study, the *Noir Marron* have conserved their African genetic ancestry, as well as their linguistic richness with an oral tradition claiming a West African origin (Fortes-Lima *et al.*, 2017). This study also established genetic relationships linking the *Noir Marron* with populations inhabiting the Bight of Benin, suggesting a genetic and linguistic correlation between these populations and their African past.

1.3. Genetic Ancestry Legacy in South America

In addition to the major role that genetic data play in elucidating disease susceptibilities, genetic data are increasingly being used to reconstruct ancestral origins and to identify familial ties, even when they extend back for hundreds of years (Underhill and Kivisild, 2007; Novembre and Ramachandran, 2011). With respect to the transatlantic African Diaspora, genetic data are proving to be important complements to historical, linguistic, ethnographic and archaeological data in these family tree reconstructions (Royal *et al.*, 2010; Stefflova *et al.*, 2011), and unexpected insights into human diseases (Winkler *et al.*, 2010).

Genetic ancestry profiles of human populations are a valuable tool to further understand the dynamics of migration and colonisation events, as well as to determine admixture patterns of populations. Historically, some of the first studies of genetic admixture at the molecular level were those that analysed the frequencies of different blood group protein alleles in African descendants, comparing them to European-Americans and Africans (Glass and Li, 1953).

Historical records on the origin of African descendants in the United States have estimated that about 64% were from West Africa, 35% from West Central Africa, and 1% from South East Africa (Thomas, 1998; McMillin, 2004). In agreement with these historical data, Salas *et al.* (2004) estimated the quantitative contribution of the different African regions to the formation of the New World mitochondrial DNA (mtDNA) gene pool. According to the estimated admixture coefficients, 28% of shared mtDNA sequence types found in North America have a West Central African origin and 72% a West African contribution. These values are significantly different from those obtained for Central America (41% West-Central and 59% West African), and South America (65% West-Central and 45% West African). Using the same kind of approach, but with substantially more data (an African database of 4,860 mtDNAs, and a database of 1,148 mtDNAs of African descendants from the U.S. that contained 1,053 mtDNAs of sub-Saharan ancestry), Salas *et al.* (2005a) estimated that >55% of the U.S. mtDNA lineages have a West African ancestry, with <41% coming from West Central or South West Africa.

Heritage is, however, too complex to be reduced to simple genetic sequences. In conjunction with data from other disciplines, genetic data can give more robust information on the African Diaspora, evidence for gene flow between Africans from various regions, as well as from non-Africans (European and Native American groups), and evidence for current genetic drift and in some cases founder effects in specific locales. For example, demographic-genetic studies among specific groups of African-Brazilians have indicated different geographical sources of the African slaves in the four major Brazilian regions (Silva *et al.*, 2006; Saloum de Neves Manta *et al.*, 2013).

In other cases, genetics, as an independent source of data, can confirm or refute existing historical reports. An example is the recent report among African-Brazilians in São Paulo (Brazil), which suggests that the relative African ancestral contributions are from West Central (45%), West (43%) and South East Africa (12%) (Goncalves *et al.*, 2008). These data agree with reports in historical

documents (Klein, 1986; Andrews, 1991). Merging such valuable genetic data with non-genetic sources (e.g., historical, archaeological and linguistic data) can yield important details and specificity to our reconstructions of the diverse events associated with the transatlantic African Diaspora and its aftermath (Jackson and Borgelin, 2010).

Thus, historical records indicate that enslaved Africans came from various regions throughout continental Africa, and with the advent of anthropological genetics, we have the opportunity to determine their specific regional ancestries. Molecular genetic studies have been used to trace African regional origins of many of their descendants, and to reconstruct the proportions of ancestry derived from different African regions (Shriver and Kittles, 2004; Ely *et al.*, 2006; Bryc *et al.*, 2010). The more varied the types of genes and gene systems that are employed in these reconstructions, the stronger the regional associations. For example, previous genetic studies retracing the genetic identity of the *Noir Marron* in Suriname and French Guiana have shown that these African-descendant communities still preserve a high conservation of the African gene pool in all their genetic systems; 99% in their mtDNA, and similar high percentages in the non-recombining region of the Y-chromosome (NRY), Gm polymorphisms of immunoglobulins, and in their human T-cell lymphotropic virus (HTLV) types (Brucato *et al.*, 2009; 2010; Fortes-Lima *et al.*, 2017).

Nevertheless, there is a significant lack of knowledge on the ancestry of African descendants in South America, especially when the disparity of markers and sampling criteria that have been used in different publications is taken into consideration (Bortolini *et al.*, 1995, 1999; Galanter *et al.*, 2012). Additional analyses using larger random samples to include new populations can be helpful in determining new aspects of the genetic structure of African descendants in Suriname, French Guiana, Colombia and Brazil (Fortes-Lima, 2016). To accomplish this objective, we performed the genotyping of different genetic markers, statistical analyses on new data and previously published population genetics databases, and comparisons of our results with those from other publications (Fortes-Lima *et al.*, 2017).

2. Discussion

2.1. Sources of African Genetic Ancestral Information in African Descendants

Since human genomes vary from individual to individual, no two unrelated individuals are alike genetically or phenotypically. With the development of various molecular techniques, the application of genetics in the study of human evolution gave rise to the fields of population genetics and molecular anthropology. Various informative and polymorphic genetic markers were discovered, and the gene frequency data emerging from their analyses contribute significantly to the successful study of evolution and worldwide human diversity (Cavalli-Sforza *et al.*, 1994; Colonna *et al.*, 2011).

The use of uniparental markers (Y-chromosome and mtDNA) and biparental markers for genetic diversity studies are promising. Y-chromosome and mtDNA provide straightforward genetic histories, due to their lack of recombination and uniparental inheritance (Underhill and Kivisild, 2007). Both accumulate mutations that can be identified by binary polymorphisms, defining haplogroups or clades for each individual or population. The frequency distribution of these genetic markers allows us to compare different human ethnic-groups through their paternal and maternal ancestry. Thus, these genetic markers are extremely powerful in characterizing the genetic history of admixture in the Americas, due to the contemporary populations of African descendants are the result of admixture between three source populations; European, Native American, and African (Parra *et al.*, 1998; Benn Torres *et al.*, 2007).

2.1.1. Role of Y-chromosome Variation in the Identification of African Origins

The Y-chromosome is inherited paternally from father to son; approximately 90% of Y-DNA is not homologous to the X-chromosome and it is termed the male specific Y-chromosome region. It contains the largest non-recombining block in the human genome, so is considered one of the most informative haplotyping systems, with important applications in evolutionary population studies, forensic, medical genetics and genealogical reconstructions (Underhill and Kivisild, 2007; Roewer, 2009).

Genetic markers on the non-recombining portion of the Y-chromosome have gradually emerged as an important tool for analysing human phylogenetic relationships. These markers represent human genetic diversity based on *Single Nucleotide Polymorphisms* (Y-SNPs) and *Short Tandem Repeats* (Y-STRs). There is now extensive knowledge on the geographic origins of Y-SNPs based on studies of global populations (Hammer *et al.*, 2001; Jobling and Tyler-Smith, 2003; Chiaroni *et al.*, 2009). The present nomenclature system of Y-chromosome genotypes has defined 20 main haplogroups, designated A through to T (Karafet *et al.*, 2008). Because of the high geographic specificity of Y-SNPs (Fortes-Lima *et al.*, 2015; Jobling and Tyler-Smith, 2003; Karafet *et al.*, 2008), Y-SNP haplogroups can be used directly to measure admixture among diverse populations without resorting to more complex models of admixture (Salas *et al.*, 2004; Mendez *et al.*, 2013). For example YAP is an ALU insertion on the NRY associated to haplogroup E, is known to occur in high frequencies within African populations and rare in non-African populations, and difference in the frequencies of the YAP insertion are indicative of varying levels of admixture across the different populations (Hammer, 1994; Torres *et al.*, 2012).

2.1.2. Role of Mitochondria DNA Variation in the Identification of African Origins

Although mtDNA represents only a small fraction of the total genome of an organism, it has emerged as one of the most popular markers to determine molecular diversity in humans over the last three decades. It has exclusive maternal inheritance, appears in multiple copies in each cell and possesses few important conserved coding sequences. In mitochondrial genomes, the mutation rate is several times higher than that of nuclear sequences. The application of mtDNA to trace evolutionary patterns and migration events in humans is based on the fact that specific haplotypes are observed in people from certain geographical regions, which might have occurred due to accumulation of mutations in different maternal lineages as people migrated (Underhill and Kivisild, 2007).

Correlations between African mtDNA haplotypes and linguistic-geographic groups show that most U.S. African Americans have mtDNA haplotypes predominantly from contemporary Niger-Congo speakers (Ely *et al.*, 2006). This finding is consistent with the historical evidence of regional sources of human trafficking during the transatlantic trade (Gomez, 2004), but many of these mtDNA variants are also observed among many Afro-Asiatic speakers and a few lineages of Nilo-Saharan and Khoe-San speakers. The extensive linguistic and geographic distributions of these identical haplotypes suggest that African women have been dispersed widely across the continent for tens of thousands of years, long before the transatlantic trade of enslaved Africans (Jackson and Borgelin, 2010). Therefore, the contemporary U.S. African-American mtDNA heterogeneity is not simply a function of more recent colonial admixture. A number of studies agree that African-descendant mtDNA haplotypes contain variation that is consistent with that described in diverse continental African populations (Salas *et al.*, 2004; Stefflova *et al.*, 2011).

According to the African-descendant forensic mtDNA database of the Scientific Working Group on DNA Analysis Methods (SWGDAM), only 8% of haplogroups observed in African descendants are more common in non-Africans, suggesting that present-day African-descendant maternal lineages significantly represent an amalgamation of African mtDNA variants with minimal non-African mtDNA variants. The most common specific haplotype among African descendants is L2a (18.8%), followed by L1c (11%), L1b (9.2%), L3e2 (9%), and L3b (8.1%); all are members of the ancient Pan-African L haplogroups (Allard *et al.*, 2005). This genetic evidence supports historical reports showing that the current African diversity observed in U.S. African-American mtDNA haplotypes was likely to be present among the earliest African captives, including the ethnic Ibo (from the Bight of Biafra region) and the ethnic Wolofs (from the Senegambia region) (Gomez, 2004).

2.2. Biological and Cultural African Heritages in Colombia

African slavery in Colombia began in the first decade of the 16th century. By the 1520s, enslaved Africans were being imported steadily into Colombia, to replace the rapidly declining Native American population. They were forced to work in gold mines, and on sugar cane plantations, cattle ranches, and large haciendas. Thus African labour was essential in different regions of Colombia, for example enslaved Africans pioneered the extracting of alluvial gold deposits and the growing of sugar cane in the states of Chocó, Antioquia, Cauca, Valle del Cauca, and Nariño in western Colombia (Navarrete, 2012).

According to the last census in Colombia (DANE, 2005), the percentage of people who self-identify as African-Colombian was around 11%; this includes being called African-Colombian, African descendant, black, mulato(a), negro(a), palenquero (African-Colombian from Palenque de San Basilio) and raizal (African-Colombian from Archipelago of San Andrés, Providencia and Santa Catalina). In total, the African-Colombian population was about 4,395,649 of the total 41,468,384 inhabitants in Colombia.

In addition, the highest presence of African-Colombians is in the Chocó region. This is the second largest population of African descendants in the Americas (after Brazil), and has been referred to as the 'African heart of Colombia'. It is important to mention the almost exclusively African-descendant community in El Palenque de San Basilio. Simply known as *'Palenque'*, it is located about 60 km inland from the city of Cartagena de Indias, and was founded in the second half of the 17th century. This former Maroon community has been speaking Spanish, and the Spanish-derived Creole language called Palenquero since its foundation (Schwegler, 2006, 2011).

Colombia has been the focus of a large number of genetic studies (Carvajal-Carmona *et al.*, 2003; Rojas *et al.*, 2010; Salas *et al.*, 2005b). Most of them, however, comprise of forensic genetic analyses (e.g., STR databases of forensic use; Paredes *et al.*, 2003), with little focus on anthropological or demographic issues. One of the first attempts to unravel the ancestry of the *'Mestizo'* or African-Colombians was performed by Rodas *et al.* (2003), by analyzing mtDNA variation. Focusing on a phylogeographic approach of mtDNA, Salas *et al.* (2008) investigated the genetic ancestry of admixed Colombian groups including *'Mestizos'* (the American term to designate individuals of European and Native American co-ancestry), *'Mulatos'* (the Colombian term to designate individuals of African and European co-ancestry), and African-Colombians. Lastly, a recent study on African-Colombians from Palenque using a Y-chromosome perspective observed a closer genetic proximity between Palenque and Cabo Verde, which can be explained by their similar European-dominated colonial history (Portuguese and Spanish slave trade), as well as their similar European and African admixture contributions (Noguera *et al.*, 2013).

2.3. Biological and Cultural African Heritages in Brazil

According to historical data, Africans who were brought to Brazil as slaves originated mainly from two geographical regions: (1) West Central and South East Africa (both former Portuguese colonies: Angola and Mozambique, respectively) and the Congo; and (2) West Africa, which covers all regions North of the Gulf of Guinea (Klein, 2002). These two major regions are large in area, and are populated by very distinct people and cultures. Most of them, however, are inhabited by speakers of languages belonging to the Niger-Congo linguistic subphylum. This subphylum comprises of the largest Bantu branch, which includes about 500 languages spoken in virtually all of Central-South Africa, except for the area occupied by the Khoe-San speaking groups (Paul *et al.*, 2013).

The contemporary Brazilian population presents the most important representative of African mtDNA lineages outside Africa. It has been estimated that at least 90 million individuals in Brazil, independent of their physical appearance, have mtDNA of sub-Saharan African origin (Pena *et al.*, 2009).

Regarding studies on mtDNA variability in Brazilian populations, the majority focus on small communities of African-descendant (Bortolini *et al.*, 1999), and only few describe the composition of urban admixed populations (Alves-Silva *et al.*, 2000). In general, lineage markers have shown that in almost all Brazilian populations studied until now, admixture was characterised by an asymmetric mating pattern occurring preferentially between European men and Native American or African women (Fortes-Lima *et al.*, 2017; Salzano, 2004; Pena *et al.*, 2009). Recently an asymmetric mating pattern was also observed between African men and Native American women in the African-Brazilian communities (*'Quilombos'*) of Maranhão in the Brazilian Amazonia (Pereira *et al.*, 2012; Saloum de Neves Manta *et al.*, 2013).

2.4. Biological and Cultural African Heritages in Suriname and French Guiana

The *Noir Marron* communities in Suriname and French Guiana provide tangible examples of how genetics can be used to answer questions on cultural and biological inheritance. The *Noir Marron* have had limited mixture with non-African, European and Native American populations, and they continue to maintain rituals inherited from West Africa, like other Maroon communities in Colombia, Jamaica and Barbados.

Brucato *et al.* (2009, 2010) investigated several aspects of the *Noir Marron* gene pool, within the context of their culture and history. They confirmed the high conservation of the African gene pool in all their genetic systems: mtDNA (99%), NRY (98%), Gm polymorphisms of immunoglobulins (97%) and human T-cell lymphotropic virus types; unique in each African-descendant group. Moreover, populations inhabiting the Bight of Benin (current Benin, Togo and West Nigeria) presented the most genetic similarities and sex-biased ancestries. It is feasible that maternal ancestry came from the Bight of Benin (Admixture rate; mY= 0.64) down to Angola (mY= 0.23) and the paternal ancestry from the Bight of Benin (mY= 0.74) up to Senegambia (mY= 0.26). The observed sex-biased admixture is supported by many historical events occurring during the slave trade (Klein, 1999; Eltis and Richardson, 2010).

2.5. *Multidisciplinary Insights into African Ancestry in South America*

In this review, we explore historical, linguistic and genetic data in order to shed light on the history of slaves in African-descendant populations from Suriname, French Guiana, Colombia and Brazil. This study is part of the EUROTAST Project (http://eurotast.eu/), an interdisciplinary research project uncovering the history and contemporary legacies of the transatlantic slave trade, through a network of thirteen research teams in history, archaeology, social anthropology and population genetics (Marchant, 2011).

For this purpose, we genotyped DNA samples of African populations from West Africa (Benin, Ivory Coast and Mali) and African descendants in South America with different colonial pasts using different genetic markers with high levels of geographic discrimination, including the whole mitochondrial genome, Y-chromosome (Y-SNPs and Y-STRs) and genome-wide SNP data (Fortes-Lima *et al.*, 2017) in order to address admixture events that have occurred many hundreds of generations ago during the slave trade, and to shed light on their African geographic origins (see Fortes-Lima *et al.*, 2016; 2017).

3. Conclusions

Recent studies have confirmed a higly preserved African genetic heritage in *Noir Marron* communities from Suriname and French Guiana, and how different is from others African descendants in Colombia and Brazil (Fortes-Lima, 2016; 2017). By combining genetic, linguistic and historical approaches not only will the ancestry of enslaved Africans be investigated in unprecedented detail, but also a whole new chapter in the identity of African descendants will be opened.

Acknowledgements

We would like to thank Hannes Schroeder, Antonio Salas, and Jay Haviser for his helpful discussions. The research leading to these results has received funding from the People Programme (Marie Curie Actions) of the European Union's Seventh Framework Programme FP7/2007–2013 under REA grant agreement n° 290344, and the French research grant; Programme Interdisciplinaire de recherche CNRS *"Amazonie - Analyse, modélisation et ingénierie des systèmes amazoniens"*. This material reflects only the author's views and the Union is not liable for any use that may be made of the information contained therein. CF-L was supported by the EUROTAST Marie Curie Initial Training Network (FP7/2007–2013).

References

Andrews, G.R. 1991. *Blacks and Whites in São Paulo Brazil 1888–1988*. The University of Wisconsin Press, Madison.

Allard, M.W., Polanskey, D., Miller, K., Wilson, M.R., Monson, K.L., Budowle, B. 2005. Characterization of human control region sequences of the African American SWGDAM forensic mtDNA data set. *Forensic Science International* 148(2-3), 169–179.

Alves-Silva, J., da Silva Santos, M., Guimaraes, P.E., Ferreira, A.C., Bandelt, H.J., Pena, S.D., Prado, V.F. 2000. The ancestry of Brazilian mtDNA lineages. *American Journal of Human Genetics* 67(2), 444–461.

Benn Torres, J., Kittles, R.A., Stone, A.C. 2007. Mitochondrial and Y Chromosome Diversity in the English-Speaking Caribbean. *Annals of Human Genetics* 71(Pt 6), 782–790.

Bortolini, M.C., Weimer, Tde, A., Salzano, F.M., Callegari-Jacques, S.M., Schneider, H., Layrisse, Z., Bonatto, S.L. 1995. Evolutionary relationships between black South American and African populations. *Human Biology* 67(4), 547–559.

Bortolini, M.C., Da Silva, W.A.J., De Guerra, D.C., Remonatto, G., Mirandola, R., Hutz, M.H., Weimer, T.A., Silva, M.C., Zago, M.A., Salzano, F.M. 1999. African-derived South American populations: A history of symmetrical and asymmetrical matings according to sex revealed by bi- and uni-parental genetic markers. *American Journal of Human Biology* 11(4), 551–563.

Brucato, N., Tortevoye, P., Plancoulaine, S., Guitard, E., Sanchez-Mazas, A., Larrouy, G., Gessain, A., Dugoujon, J.M. 2009. The genetic diversity of three peculiar populations descending from the slave trade: Gm study of Noir Marron from French Guiana. *Comptes Rendus Biologies* 332(10), 917–926.

Brucato, N., Cassar, O., Tonasso, L., Tortevoye, P., Migot-Nabias, F., Plancoulaine, S., Guitard, E., Larrouy, G., Gessain, A., Dugoujon, J.M. 2010. The imprint of the Slave Trade in an African American population: mitochondrial DNA, Y chromosome and HTLV-1 analysis in the Noir Marron of French Guiana. *BMC Evolutionary Biology* 10, 314.

Bryc, K., Auton, A., Nelson, M.R., Oksenberg, J.R., Hauser, S.L., Williams, S., Froment, A., Bodo, J.M., Wambebe, C., Tishkoff, S.A., Bustamante, C.D. 2010. Genome-wide patterns of population structure and admixture in West Africans and African Americans. *Proceedings of the National Academy of Sciences of the United States of America* 107(2), 786–791.

Carvajal-Carmona, L.G., Ophoff, R., Service, S., Hartiala, J., Molina, J., Leon, P., Ospina, J., Bedoya, G., Freimer, N., Ruiz-Linares, A. 2003. Genetic demography of Antioquia (Colombia) and the Central Valley of Costa Rica. *Human Genetics* 112(5-6), 534–541.

Cavalli-Sforza, L.L., Menozzi, P., Piazza, A. 1994. *The History and Geography of Human Genes*. Princeton University Press, Princeton.

Colonna, V., Pagani, L., Xue, Y., Tyler-Smith, C. 2011. A world in a grain of sand: Human history from genetic data. *Genome Biology* 12(11), 234.

Chiaroni, J., Underhill, P.A., Cavalli-Sforza, L.L. 2009. Y chromosome diversity, human expansion, drift, and cultural evolution. *Proceedings of the National Academy of Sciences of the United States of America* 106(48), 20174–20179.

DANE, Departamento Administrativo Nacional de Estadística, 2005. Censo General 2005: http://www.dane.gov.co/index.php/es/poblacion-y-registros-vitales/censos/censo-2005 (accessed 30th November, 2013).

Eltis, D. 2009. Voyages: The Trans-Atlantic Slave Trade Database. http://www.slavevoyages.org (accessed 30th November, 2013).

Eltis, D., Richardson, D. 2010. *Atlas of the Transatlantic Slave Trade. The Lewis Walpole Series in Eighteenth-Century Culture and History*. Yale University Press, New Haven.

Ely, B., Wilson, J.L., Jackson, F., Jackson, B.A. 2006. African-American mitochondrial DNAs often match mtDNAs found in multiple African ethnic groups. *BMC Biology* 4, 34.

EUROTAST Initial Training Network. Exploring the history, archaeology and new genetics of the Trans-Atlantic Slave Trade. http://eurotast.eu/ (accessed 30th November, 2013).

Fortes-Lima, C. 2016. *Tracing the genetic origin of African descendants from South America*. Editions Universitaires Européennes, Germany.

Fortes-Lima, C., Brucato, N., Croze, M., Bellis, G., Schiavinato, S., Massougbodji, A., Migot-Nabias, F., Dugoujon, JM. 2015. Genetic population study of Y-chromosome markers in Benin and Ivory Coast ethnic groups. *Forensic Science International Genetics* 19(1), 232–237.

Fortes-Lima, C., Gessain, A., Ruiz-Linares, A., Bortolini, M.C., Migot-Nabias, F., Bellis, G., Moreno-Mayar, J.V., Restrepo, B.N., Rojas, W., Avendaño-Tamayo, E., Bedoya, G., Orlando, L., Salas, A., Helgason, A., Gilbert, M.T.P., Sikora, M., Schroeder, H., Dugoujon, J.M. 2017. Genome-wide Ancestry and Demographic History of African-Descendant Maroon Communities from French Guiana and Suriname. *American Journal of Human Genetics* 101(5), 725–736.

Galanter, J.M., Fernandez-Lopez, J.C., Gignoux, C.R., Barnholtz-Sloan, J., Fernandez-Rozadilla, C., Via, M., Hidalgo-Miranda, A., Contreras, A.V., Figueroa, L.U., Raska, P., Jimenez-Sanchez, G., Zolezzi, I.S., Torres, M., Ponte, C.R., Ruiz, Y., Salas, A., Nguyen, E., Eng, C., Borjas, L., Zabala, W., Barreto, G., Gonzalez, F.R., Ibarra, A., Taboada, P., Porras, L., Moreno, F., Bigham, A., Gutierrez, G., Brutsaert, T., Leon-Velarde, F., Moore, L.G., Vargas, E., Cruz, M., Escobedo, J., Rodriguez-Santana, J., Rodriguez-Cintron, W., Chapela, R., Ford, J.G., Bustamante, C., Seminara, D., Shriver, M., Ziv, E., Burchard, E.G., Haile, R., Parra, E., Carracedo, A., LACE Consortium, 2012. Development of a panel of genome-wide ancestry informative markers to study admixture throughout the Americas. *PLoS Genetics* 8(3), e1002554.

Glass, B., Li, C.C. 1953. The dynamics of racial intermixture; an analysis based on the American Negro. *American Journal of Human Genetics* 5(1), 1–20.

Gomez, M.A. 2004. *Exchanging Our Country Marks: Transformation of African Identities in the Colonial and Antebellum South*. University of North Carolina Press, Chapel Hill.

Goncalves, V.F., Carvalho, C.M., Bortolini, M.C., Bydlowski, S.P., Pena, S.D. 2008. The phylogeography of African Brazilians. *Human Heredity* 65(1), 23–32.

Hammer, M.F. 1994. A recent insertion of an Alu element on the Y chromosome is a useful marker for human population studies. *Molecular Biology and Evolution* 11(5), 749–761.

Hammer, M.F., Karafet, T.M., Redd, A.J., Jarjanazi, H., Santachiara-Benerecetti, S., Soodyall, H., Zegura, S.L. 2001. Hierarchical patterns of global human Y-chromosome diversity. *Genome Biology and Evolution* 18(7), 1189–1203.

Huttar, G.L. 2009. Development of a creole lexicon, in: Selbach, R., Cardoso, H.C., van den Berg, M. (Eds.), *Gradual Creolization: Studies Celebrating Jacques Arends*. John Benjamins Publishing Company, Amsterdam, pp. 173–188.

Huttar, G.L. 2010. Successive African and European contributions to suriname Creole lexicons, in: McElhanon, K.A., Reesink,

G. (Eds.), *A Mosaic of Languages and Cultures*. SIL International, Dallas, pp. 180–191.

Huttar, G.L. 2012. The African lexical contribution to Ndyuka, Saramaccan, and other creoles: Implications for how creoles develop, in: Bartens, A., Baker, P. (Eds.), *Black through White: African Words and Calques which Survived Slavery in Creoles and Transplanted European Languages*, Westminster Creolistics Series. Battlebridge Publications, London, pp. 215–229.

Jackson, F.L.C., Borgelin L.F.J. 2010. How genetics can provide detail to the transatlantic African diaspora, in: Olaniyan, T., Sweet, J.H. (Eds.), *The African Diaspora and the Disciplines*. Indiana University Press, Bloomington, pp. 75–100.

Jobling, M.A., Tyler-Smith, C. 2003. The human Y chromosome: An evolutionary marker comes of age. *Nature Review Genetics* 4(8), 598–612.

Karafet, T.M., Mendez, F.L., Meilerman, M.B., Underhill, P.A., Zegura, S.L., Hammer, M.F. 2008. New binary polymorphisms reshape and increase resolution of the human Y chromosomal haplogroup tree. *Genome Research* 18(5), 830–838.

Klein, H.S. 1986. *African Slavery in Latin America and the Caribbean*. Oxford University Press, New York.

Klein, H.S. 1999. *The Atlantic Slave Trade*. Cambridge University Press, Cambridge.

Klein, H.S. 2002. As origens africanas dos escravos brasileiros, in: Pena, S.D.J. (Ed.), *Homo brasilis. Aspectos genéticos, lingüisticos, históricos e socioantropológicos da formação do povo brasileiro*. FUNPEC Editora, Ribeirão Preto, pp. 93–112.

Lefebvre, C. 2013. A comparison of the nominal structures of Saramaccan, Fongbe and English with reference to Haitian Creole: Implications for a relabelling-based account of creole genesis. *Lingua*, 129.

Marchant, J. 2011. Filling in the gaps in the slave trade. *Nature News*. doi:10.1038/nature.2011.9535.

McMillin, J.A. 2004. *The Final Victims: Foreign Slave Trade to North America 1783–1810*. University of South Carolina Press, Columbia.

Mendez, F.L., Krahn, T., Schrack, B., Krahn, A.M., Veeramah, K.R., Woerner, A.E., Fomine, F.L., Bradman, N., Thomas, M.G., Karafet, T.M., Hammer, M.F. 2013. An African American paternal lineage adds an extremely ancient root to the human Y chromosome phylogenetic tree. *American Journal of Human Genetics* 92(3), 454–459.

Navarrete, M.C. 2012. Palenques: cimarrones y castas en el Caribe colombiano – Sus relaciones sociales (siglo XVII), in: Maglia, G., Schwegler, A. (Eds.), *Palenque (Colombia): oralidad, identidad y resistencia. Un enfoque interdisciplinario*. Instituto Caro y Cuervo & Universidad Javeriana, Bogota, pp. 257–284.

Nelson, A. 2016. *The social life of DNA: Race, reparations, and reconciliation after the genome*. Beacon Press, Boston.

Noguera, M.C., Schwegler, A., Gomes, V., Briceno, I., Alvarez, L., Uricoechea, D., Amorim, A., Benavides, E., Silvera, C., Charris, M., Bernal, J.E., Gusmao, L. 2013. Colombia's racial crucible: Y chromosome evidence from six admixed communities in the Department of Bolivar. *Annals of Human Biology*. doi:10.3109/03014460.2013.852244.

Novembre, J., Ramachandran, S. 2011. Perspectives on human population structure at the cusp of the sequencing era. *Annual Review of Genomics and Human Genetics* 12, 245–274.

Paredes, M., Galindo, A., Bernal, M., Avila, S., Andrade, D., Vergara, C., Rincon, M., Romero, R.E., Navarrete, M., Cardenas, M., Ortega, J., Suarez, D., Cifuentes, A., Salas,

A., Carracedo, A. 2003. Analysis of the CODIS autosomal STR loci in four main Colombian regions. *Forensic Science International* 137(1), 67–73.

Parra, E.J., Marcini, A., Akey, J., Martinson, J., Batzer, M.A., Cooper, R., Forrester, T., Allison, D.B., Deka, R., Ferrell, R.E., Shriver, M.D. 1998. Estimating African American admixture proportions by use of population-specific alleles. *American Journal of Human Genetics* 63(6), 1839–1851.

Paul, L.M., Gary, F., Simons, G.F., Fennig, C.D. 2013. Ethnologue: Languages of the World, Seventeenth edition. Dallas, Texas: SIL International. Online version: http://www.ethnologue.com (accessed 30th November, 2013).

Pena, S.D., Bastos-Rodrigues, L., Pimenta, J.R., Bydlowski, S.P. 2009. DNA tests probe the genomic ancestry of Brazilians. *Brazilian Journal of Medical and Biological Research* 42(10), 870–876.

Pereira, R., Phillips, C., Pinto, N., Santos, C., dos Santos, S.E., Amorim, A., Carracedo, A., Gusmao, L. 2012. Straightforward inference of ancestry and admixture proportions through ancestry-informative insertion deletion multiplexing. *PLoS ONE* 7(1), e29684.

Price, R., 1996. *Maroon Societies: Rebel Slave Communities in the Americas*. John Hopkins University Press, Baltimore.

Price, R. 2002. Maroons in Surinam and Guyane: How many and where. *Nieuwe West-Indische Gids* 76, 81–88.

Reiner, A.P., Ziv, E., Lind, D.L., Nievergelt, C.M., Schork, N.J., Cummings, S.R., Phong, A., Burchard, E.G., Harris, T.B., Psaty, B.M., Kwok, P.Y. 2005. Population structure, admixture, and aging-related phenotypes in African American adults: The Cardiovascular Health Study. *American Journal of Human Genetics* 76(3), 463–477.

Rodas, C., Gelvez, N., Keyeux, G. 2003. Mitochondrial DNA studies show asymmetrical Amerindian admixture in Afro-Colombian and Mestizo populations. *Human Biology* 75(1), 13–30.

Roewer, L. 2009. Y chromosome STR typing in crime casework. *Forensic Science, Medicine and Pathology* 5(2) 77–84.

Rojas, W., Parra, M.V., Campo, O., Caro, M.A., Lopera, J.G., Arias, W., Duque, C., Naranjo, A., García, J., Vergara, C., Lopera, J., Hernandez, E., Valencia, A., Caicedo, Y., Cuartas, M., Gutiérrez, J., Lopez, S., Ruiz-Linares, A., Bedoya, G. 2010. Genetic make up and structure of Colombian populations by means of uniparental and biparental DNA markers. *American Journal of Physical Anthropology* 143(1), 13–20.

Royal, C.D., Novembre, J., Fullerton, S.M., Goldstein, D.B., Long, J.C., Bamshad, M.J., Clark, A.G. 2010. Inferring genetic ancestry: Opportunities, challenges, and implications. *American Journal of Human Genetics* 86(5), 661-673.

Salas, A., Richards, M., Lareu, M.V., Scozzari, R., Coppa, A., Torroni, A., Macaulay, V., Carracedo, A. 2004. The African diaspora: Mitochondrial DNA and the Atlantic slave trade. *American Journal of Human Genetics* 74(3), 454–465.

Salas, A., Carracedo, A., Richards, M., Macaulay, V. 2005a. Charting the ancestry of African Americans. *American Journal of Human Genetics* 77(4), 676–680.

Salas, A., Richards, M., Lareu, M.V., Sobrino, B., Silva, S., Matamoros, M., Macaulay, V., Carracedo, A. 2005b. Shipwrecks and founder effects: Divergent demographic histories reflected in Caribbean mtDNA. *American Journal of Physical Anthropology* 128(4), 855–860.

Salas, A., Acosta, A., Alvarez-Iglesias, V., Cerezo, M., Phillips, C., Lareu, M.V., Carracedo, A. 2008. The mtDNA ancestry of admixed Colombian populations. *American Journal of Human Biology* 20(5), 584–591.

Saloum de Neves Manta, F., Pereira, R., Vianna, R., Rodolfo Beuttenmuller de Araujo, A., Leite Goes Gitai, D., Aparecida da Silva, D., de Vargas Wolfgramm, E., da Mota Pontes, I., Ivan Aguiar, J., Ozorio Moraes, M., Fagundes de Carvalho, E., Gusmao, L., 2013. Revisiting the genetic ancestry of Brazilians using autosomal AIM-Indels. *PLoS ONE* 8(9), e75145.

Salzano, F.M. 2004. Interethnic variability and admixture in Latin America – social implications. *Revista de Biologica Tropica* 52(3), 405–415.

Schwegler, A. 2006. Bantu elements in Palenque (Colombia), in: Haviser, J.B., MacDonald, K.C. (Eds.), *African Re-genesis: Confronting Social Issues in the Diaspora*. UCL Press, Abingdon, pp. 204–223.

Schwegler, A. 2011. Palenque-ro. The search for its African substrate, in: Lefebvre, C. (Eds.), *Creoles, their Substrates, and Language Typology*. John Benjamins Publishing Company, Amsterdam, pp. 225–249.

Shriver, M.D., Kittles, R.A. 2004. Genetic ancestry and the search for personalized genetic histories. *Nature Review Genetics* 5(8), 611–618.

Silva, W.A., Bortolini, M.C., Schneider, M.P., Marrero, A., Elion, J., Krishnamoorthy, R., Zago, M.A. 2006. MtDNA haplogroup analysis of black Brazilian and sub-Saharan populations: implications for the Atlantic slave trade. *Human Biology* 78(1), 29–41.

Smith, N. 2002. The history of the Surinamese creoles II: Origin and differentiation, in: Carlin, E.B., Arends, J. (Eds.), *Atlas of the Languages of Suriname*. The Royal Netherlands Institute of Southeast Asian and Caribbean Studies, Leiden, pp. 131–151.

Stefflova, K., Dulik, M.C., Barnholtz-Sloan, J.S., Pai, A.A., Walker, A.H., Rebbeck, T.R. 2011. Dissecting the within-Africa ancestry of populations of African descent in the Americas. *PLoS ONE* 6(1), e14495.

Thomas, H. 1998. *The Slave Trade - The History of the Atlantic Slave Trade: 1440–1870*. Macmillan, London.

Torres, J.B., Doura, M.B., Keita, S.O., Kittles, R.A. 2012. Y chromosome lineages in men of West African descent. *PLoS ONE* 7(1), e29687.

Underhill, P.A., Kivisild, T. 2007. Use of Y chromosome and mitochondrial DNA population structure in tracing human migrations. *Annual Review in Genetics* 41, 539–564.

Winkler, C.A., Nelson, G.W., Smith, M.W. 2010. Admixture mapping comes of age. *Annual Review of Genomics and Human Genetics* 11, 65–89.

3

The 'Invisible Dead' Project: The Database as a Work-in-Progress

Mandy Jay, Chris Scarre, Charlotte Roberts, Graham Philip, Jennie Bradbury and Douglas Davies

Low numbers of prehistoric burials in the archaeological record may suggest that formal burial was sometimes the exception rather than the rule. By examining archaeological data from across two regions (Britain and the Levant) this project aims to chart changing concepts of what it means to be human. One of our objectives is to examine whether burials represented in these regions constitute the 'mainstream' or whether they are the result of highly specific selection processes, i.e. whether the archaeological burial record is sparse because we have found only a small proportion of the original presence, or because only particular people were chosen for formal and visible processes.

The first step for this project is to compile a database of burial information from the two regions covering the Neolithic through to the Roman period. These data will be linked to a Geographic Information System (GIS), allowing variations in mortuary practices over time and space to be detected and analysed. Proxy data for estimated populations can then be compared, allowing us to consider how differences in the treatment of the majority of the population may have changed over time. In this way the 'Invisible Dead' may start to become visible.

Keywords Burial; Geographic Information System; Britain; Levant; Late Prehistory; Roman

1. Introduction

'The 'Invisible Dead' and the Development of Early Human Beliefs about the Body' project aims to look at the archaeological evidence for disposal of the dead in two regions of the world, Britain and the Levant, from the Neolithic through to the Roman period (4000 BC to AD 400 in Britain; 10,000 BC to AD 400 in the Levant). The data collected will ultimately provide a better understanding of burial rites in the past and one of the key issues we hope to address is the extent to which 'formal burial' can be seen as a normative mortuary practice. What proportion of the population at any given point in time or space are likely to have received a formal burial and how is this reflected in the archaeological record? During some periods, archaeological evidence for burial is particularly sparse, hence the title of the project. For these apparent 'gaps' in the record it is pertinent to ask whether the attitude to formal burial had altered, so that a more 'invisible' disposal rite was used, or whether there are other explanations available, such as biases in

excavation and archaeological recording methods or a reduction in population numbers.

Whilst this paper concentrates on the British element of the project, many of the issues and themes discussed here are also relevant for the Levant. The two regions were specifically selected for both their similarities and their differences. They lie at the eastern and western ends of the Christian world, yet by the 4th century AD both included burials sharing features that later became characteristic of Christian (and in due course Islamic) burials. The contrasting timings and nature of socio-economic developments within the two regions will allow the cross-comparison of similar burial traditions within different socio-political settings.

The biological anthropology community in Britain does not currently have quantified data for the number of individuals deposited in archaeological contexts for the overall period being studied for this project, or for burial site numbers. The data currently available for Britain only cover fragments of that remit. There are smaller collections of data available for specific periods, geographical locations, types of burial feature or particular purposes, for example Neolithic long barrows for the Cotswolds region (Darvill, 2004), Iron Age burials (Whimster, 1981), later prehistoric burials in southern Britain (Bristow, 2004), and the BABAO remit to produce a database of curated remains, but it is very difficult to bring these together to obtain an overall picture. Thus, one of the main objectives of the project is to build a database which will help us to better understand changes and patterns in the burial and skeletal record.

2. The Database

The database is an ambitious project that cannot be exhaustive in its first stage, but plans are in place to build on it in future phases. Its framework was originally designed for the Fragile Crescent Project (Durham University) and uses a combination of Microsoft Excel® and Access® for data entry, with spatial information being stored within ArcGIS®. It has been designed with flexibility in mind, allowing additions and further observations to be added over time. Entries for a burial or burial site can vary from a place name, minimum number of individuals (MNI) and geographical locale to much more detailed observations describing or categorising the skeletal remains, burial context and forms of associated material culture.

At the time of writing, the project has been running for only one year,[1] so it is a work-in-progress, but currently the British material has around 2000 sites entered, with over 30,000 observations and just under 16,000 as a minimum number of individuals. The aim has been to produce a robust prototype with the vision to expand, using a wide variety of specific terms and numerical fields which can be manipulated for analysis, rather than to focus on a restricted set of precision data (as has been done previously in lists such as those mentioned in the Introduction), or mainly text-based fields, which are available from more general databases of archaeological data such as Pastscape (English Heritage).

Some of the sites have already been entered with a lot of detail, so that we can understand how those data may be used, how they might spark questions to be answered in the future and how we might successfully extrapolate from regionally focused data. In contrast other sites have currently only been entered at a bare minimum of a map reference, site name, period code and a minimum number of individuals, this forming a 'skeleton' which can be built upon later.

For overall coverage it is important to avoid biases with respect to the type of material included. This means that all records of human remains have been entered, ranging from large cemeteries (e.g., the Eastern Cemetery of Roman London, where there are nearly 700 formal burials and a detailed publication of the osteological remains [Barber and Bowsher, 2000]), through to single isolated fragments of human bone (e.g., the distal end of a femur and a second phalanx found in a ditch terminal

at the Neolithic Stones of Stenness on Orkney [Ritchie, 1978]). Information about burial contexts which do not have a record of skeletal remains are also being included, as are data where such remains are mentioned as being present, but without further detail (e.g., in antiquarian records). This means the inclusion of monuments such as barrows which have not been excavated, or which are believed to have been present as burial contexts, but have been ploughed out (e.g., large Iron Age 'square' barrow cemeteries in East Yorkshire, which are recorded only through aerial photography, but are believed to contain many hundreds of burial sites [Stead, 1976]), through to sites where the soil properties mean that the skeletal remains are probably non-surviving (e.g., the Iron Age Newbridge 'chariot' burial, the only such British vehicle known outside of Yorkshire, but where the acid Scottish soils have left no trace of bone [Carter *et al.,* 2010]).

The words 'believed' and 'probably' were used in the last paragraph and, as with all archaeological datasets, levels of uncertainty will be attached to many of these data. The database format allows the use of certainty fields to assign a probable or possible value to an entry, both in terms of the entry itself and also in terms of the period assigned to the entry. For instance, when John Mortimer excavated barrow C58 in East Yorkshire in 1873 he described 'under the centre of the barrow' in a hole 'a few potsherds and what seemed to be a little powdered bone ash, but nothing further to indicate its age or use' (Mortimer, 1905: 263). This may be a human cremation, associated with a vessel, which may be of similar date to the other material associated with the barrow, but the certainty attached to any of these observations needs to be defined in order to allow proper comparison with other, more definite records of such evidence. In this case, an observation of a *possible* cremation, alongside a *possible* cremation urn, with a *possible* date of Late Bronze Age or Early Iron Age, has been recorded. This issue is particularly important as some gazetteers, databases and lists which have been compiled in the past have specifically omitted uncertain data. Such an approach means that many data which are difficult to quantify or categorise, such as antiquarian reports, are not included in considerations of mortuary evidence. When looking for the 'invisible' components of the record, such as the period in the Late Bronze Age and Early Iron Age where British evidence is sparse, it is clearly of concern that all avenues of potential evidence are investigated.

Numbers of individuals may also need to be estimated. Where human remains are recorded as present, without further detail, they are recorded as at least one individual being present, associated with the term 'Presence'; where there are multiple individuals, but without specific numbers available, they are recorded as a minimum of two associated with the term 'Multiple', with similar recording for 'Several' and 'Approximate'. This allows estimates of numbers to be obtained with the ability to break the data down into those which are minimum limits and those which are much more precise.

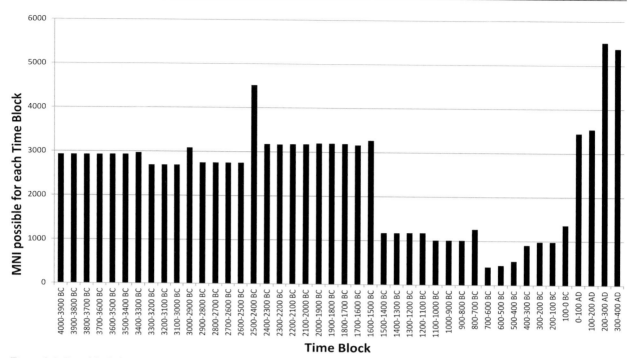

Figure 3.1 Time block distribution of MNI recorded in the database at time of writing, according to period codes. The plot is provided in 100-year time slots along the horizontal axis. The vertical axis indicates the MNI which could be present for any particular time block (see text for further discussion).

Chronology is important and again presents challenges which are fundamental to the nature of archaeology. Recorded dating may range from a recent and relatively precise radiocarbon date on an individual skeleton through to an estimate that a site is 'prehistoric' from an antiquarian report. A range of period codes with defined ranges have been used to attach to individual parts of a site record and radiocarbon dates can be entered for individual materials, but using these varying levels of precision to produce charts of changes over time is challenging. A 'prehistoric' individual may date to the same period as one with a radiocarbon date, but equally well may not.

Figure 3.1 provides an example of one of the approaches available with this database, in which time blocks can be plotted for 100-year slots. The vertical axis does not represent cumulative data because the analysis allocates them to any 100-year block to which they *could* relate, based on the time block coding. So, for instance, the period code for the Neolithic generally covers the whole of the period 4000 BC to 2400 BC, so that any site with an MNI allocated to it and a period code entered as Neolithic will have that MNI plotted in Figure 3.1 for all of the 100-year blocks in that period range, because the individuals might date to any one of those blocks. More precise period codes (e.g., Late Neolithic) will have the MNI plotted in lower numbers of time blocks. This approach can cause 'spikes' in the data (e.g., in the 2500 to 2400 BC time block in Figure 3.1) which are an artefact of the process. They are caused by the fact that the period codes can overlap, reflecting the inherent problems involved in allocating relative date brackets to archaeological material when more precise data are unavailable.

This time block approach is one of the possible ways of manipulating all of the available data without having to remove the more imprecisely dated material and can be compared with more precise analyses (e.g., charting only radiocarbon dated material). A fuller discussion of many of the approaches and of the chronological issues generally can be found in Lawrence *et al.* (2012), Bradbury *et al.* (2016) and Jay and Scarre (2017).

3. Using GIS

The project aims to analyse the data from the database both temporally and spatially. With this in mind, spatial information is being collated into a GIS which can be manipulated and analysed in programmes such as ArcGIS to produce maps for a multitude of variables. Figure 3.2 shows all of the British sites which, at the date of writing, have been entered with a grid reference and a site name.

Given the diffuse nature of the data entry to date and the variety of material plotted, this map might, in theory, be considered of little use, except as a record of progress. However, there are spatial elements to the record which are of interest. Dense plots of sites are visible in southern England, in the Wiltshire area around Stonehenge, and in East Yorkshire. These are areas with long records of focused excavation where individual excavator interests, the publication records of institutions or societies and the interest of the general public in specific monuments all come together to increase the visible presence. There are also issues of local environment and preservation to bear in mind; they are both areas of chalk where skeletal remains will preserve relatively well (although they

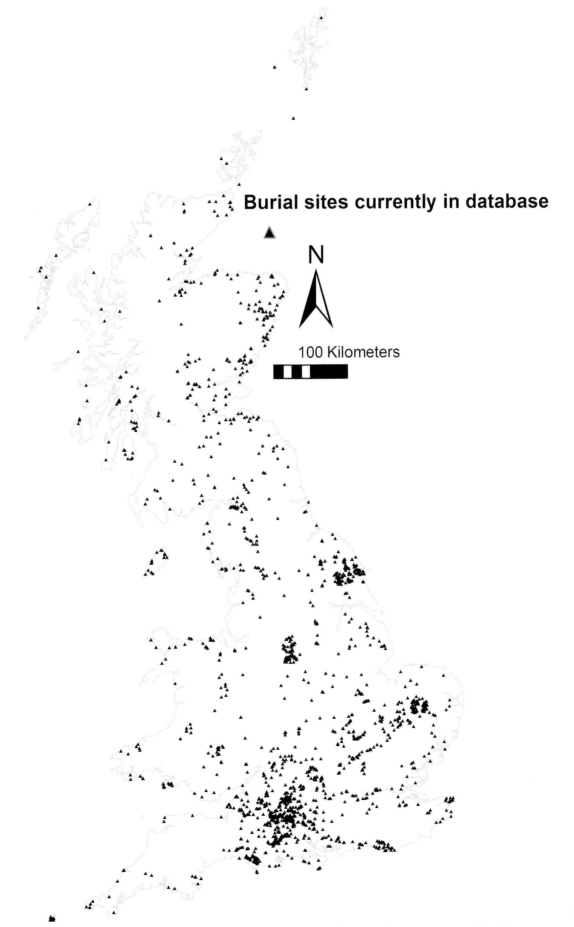

Figure 3.2 Sites recorded in the database at the time of writing, covering a wide range of dates, locations and burial context types.

may sometimes become friable, due to the permeable nature of chalk soils), and areas where subsistence would have been relatively easy during later prehistory (Manby, 1980; Brothwell, 1981: 7; Green, 2000: 36–49; Allen and Gardiner, 2009). It is to be expected that the presence in these areas would, for instance, vastly exceed central Scotland, where bone preservation in acid soils and subsistence conditions (particularly in the central mountainous areas) are usually very poor (Nielsen-Marsh *et al.,* 2007; Dobbie *et al.,* 2011). So in this respect, the pattern might be considered obvious.

There is a dense patch of sites in central England, in the Peak District, which is reflective more of the data entry pattern than of the other biases mentioned above. This relates to Late Neolithic and Early Bronze Age sites which have been entered from another database compiled by one of the authors for the Beaker People Project (Jay *et al.,* 2012; Parker Pearson *et al.,* in prep). This example illustrates that some of the patterns in the plot will, currently, be due to data entry processes which involve the use of existing lists, gazetteers and databases.

There are, therefore, biases known to affect a map such as this, but there are also interesting factors which bear consideration for the future, even at this scale. For instance, Figure 3.3 shows an apparently linear pattern of sites running along a geological escarpment. That pattern is also clearly visible on Figure 3.2 and it is not the result of a particular pattern of data entry, as was the case for the Peak District cluster mentioned above. These sites are all from different periods, from Neolithic through to Roman, and they cover a wide range of site types, from a cremation cemetery to a chambered long barrow. This arrangement of sites, therefore, does not relate to one particular type of site or to one archaeological period. Given the linearity, one explanation might be that it has been caused by a road construction corridor. Plotting the sites onto a road map (not shown) demonstrates that, whilst some of the sites may be related to specific road construction projects, this is not the explanation for the pattern as a whole. An additional form of evidence to support this can be found in the data themselves. The database contains information about excavation dates, and when these are extracted it shows that the sites in this area were excavated at different times over at least 150 years, from the 1850s through to very recently. It is unlikely, therefore, that this arrangement of sites is a product of any particular excavation/recording process.

When the sites are plotted against an Ordnance Survey map (not shown), there is strong correlation between the distribution of sites and the Icknield Way, supposedly the oldest long-distance path in Britain, going from the Chilterns up into Norfolk following a chalk escarpment (Dyer, 1963). It is possible, therefore, to match the sites to the trackway, but then there are questions raised about whether the path is as ancient as some people have suggested (Harrison, 2003), so the relationship between path and burial evidence may not be straightforward.

Figure 3.4 shows a plot of Natural England's data for ancient woodland, which is defined as woodland which dates back to at least AD 1600 (Natural England, 2002). Whilst this coverage is not necessarily analogous to prehistoric woodland, it may give some indication of landscape patterns. Here, the ridge sites run along the line where there is very little woodland to the north and quite a lot to the south. There is also a gap in the sites generally where very few are plotted in the area of East Sussex and southern Kent which has the densest plot of ancient woodland. This may provide information about the relationship between local environmental conditions and the placement of burials during the period under study, or it might also suggest that some environmental variables will affect where sites will be discovered or excavated. Historically heavily wooded areas where ploughing is less prevalent will, for instance, reduce the number of barrow sites which are recorded from aerial photographs identifying crop marks. It is clearly important to be careful of being simplistic with these apparent relationships, but these kinds of spatial analyses may well provide insights into understanding *where* the dead are 'invisible' as well as *when*.

Figure 3.5 shows the period for which we might most expect to encounter the 'Invisible Dead' in Britain. It plots the sites where evidence is dated to the Late Bronze Age or Early Iron Age, when the record is much sparser than for other periods (Brück, 1995). Burial monuments, such as barrows, are rarely identified for this period (Champion, 2009), so that this reduction in overall presence may relate partly to our ability to identify the sites where burials occur, alongside other factors, such as the possibilities that formal mortuary practices (as we would recognise them) were less important during this period, or that less visible deposition methods (e.g., scattered cremations) were the norm. Bearing in mind that this dataset is not complete at this stage of the work, data for this period so far have been classified on this map according to the Minimum Number of Individuals recorded for the site, and the sites which were recorded as only *possibly* of the relevant date have been marked separately. This gives just one example of how the data can be visually manipulated to any one of a large number of possibilities. Other examples might be to map only those sites at which males were identified, or only those sites where radiocarbon dates on skeletal material were available.

Even as a work in progress, the data for this period are showing a distribution which is reminiscent of the overall dataset in Figure 3.2. There are clusters in the south and in eastern Yorkshire, and there are also gaps in the distribution where the more wooded areas from the ancient woodland map occur. Although the numbers are much lower for this period than they are at other times (see Figure 3.1), they are presenting in a similar spatial way. This suggests that the factors affecting spatial distribution of identified burials may not have changed during this period, even if the numbers do. Ultimately, as the work continues, this will have implications when discussing issues such as the

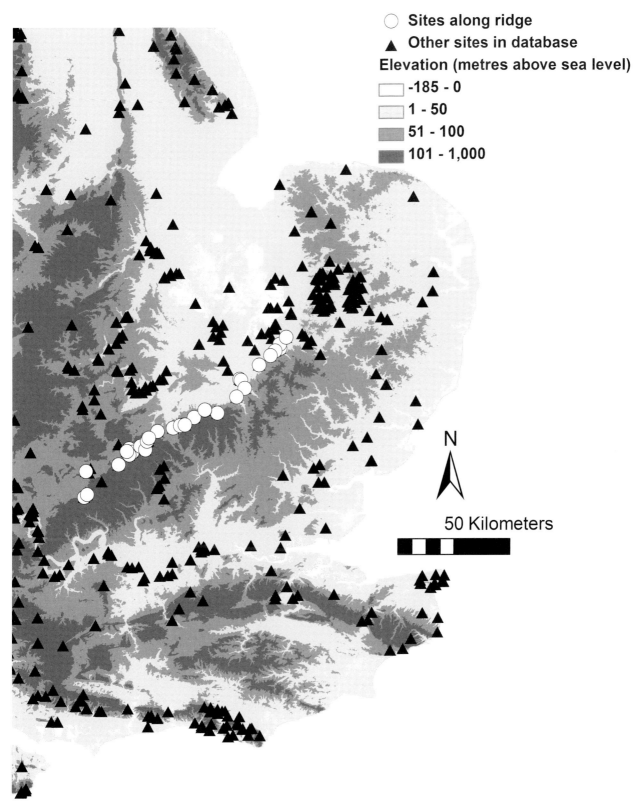

Figure 3.3 Sites highlighted in black located along the line of a natural chalk escarpment which is followed by the Icknield Way.

contraction of human land use during this period due to environmental change and human activity, or the patchy nature of the record being affected by uneven coverage of archaeological observation (Champion, 2009).

Early indications are that at least some of the 'missing' individuals for this part of the Bronze Age and Early Iron Age are becoming more visible with recent work. This is partly due to more extensive radiocarbon dating programmes and the direct dating of cremations which is a relatively recent advance, although the precision of these needs to be considered with care (Lanting *et al.*, 2001; Zazzo *et al.*, 2012), but it is also a product of a

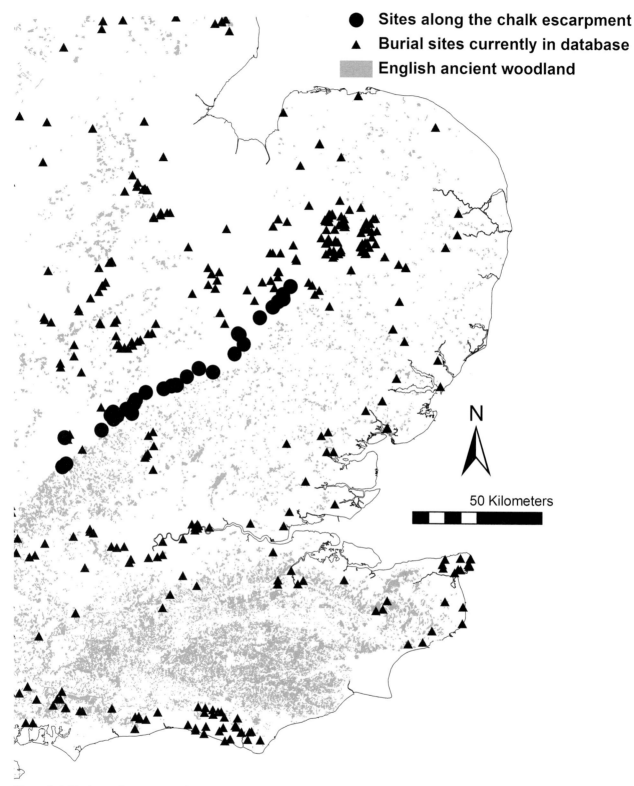

Figure 3.4 Shading indicates Natural England data for Ancient Woodland (© Natural England), defined as woodland present at least as far back as AD 1600. Sites highlighted in black are those along the line of the escarpment in Figure 3.3.

willingness to question the assumed date of a context. Whilst burial monuments appear to have waned at this time, it is clear that the earlier barrows attracted intrusive burials which were often inserted a long time after the original construction. Burial locations attract burials throughout not only the period being investigated, but

far beyond. In one exceptional case, the English Heritage Pastscape database records that there is evidence for a 20th-century druid cremation burial in a Bronze Age barrow on Normanton Down, near Stonehenge (English Heritage, Pastscape monument no. 943053). Such intrusive burials tend to be relatively easily identified

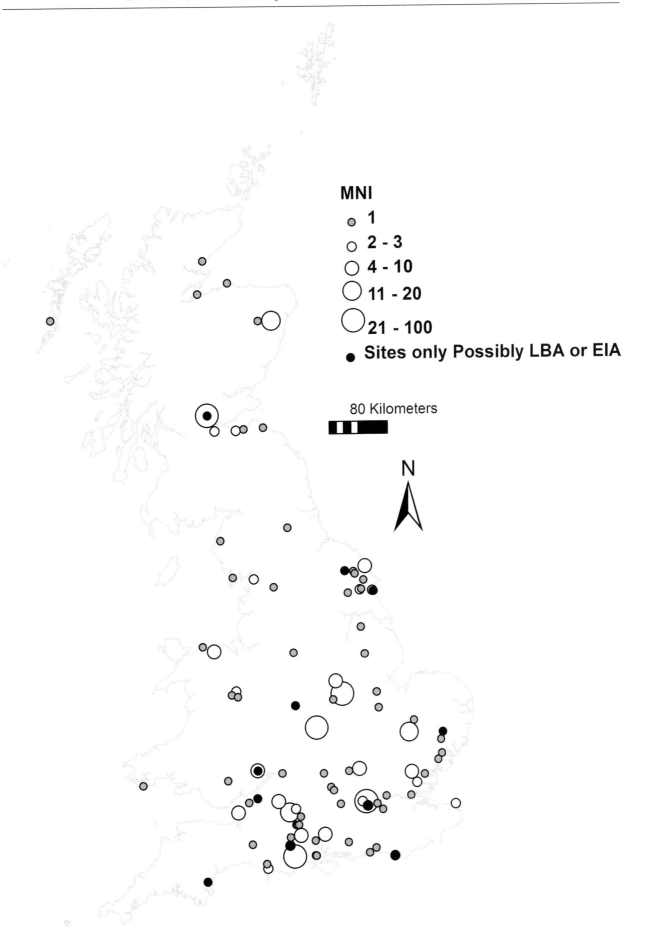

Figure 3.5 Minimum Number of Individuals plotted for sites which are dated to the Late Bronze Age or Early Iron Age, graduated symbols indicating the number of individuals for each site, with those sites which have been designated as only 'possible' for this period indicated separately.

as non-prehistoric when there are artefacts present and this often allows Anglo-Saxon insertions, for instance, to be dated. However, where no such objects are present, the dating of such material has sometimes been assumed to be closer to the period of the barrow itself, or else no date is attributed at all. In some cases, radiocarbon dating shows that some of the 'invisible' material may be found in this way, so that identifying barrows which may have been used as a focus for burials over a long period of time may help us to understand what is happening. An example of such a burial has been dated from an Early Bronze Age barrow at Garton Slack in East Yorkshire as part of the Beaker People Project. Originally thought to have been associated with an Early Bronze Age jet button, the radiocarbon date for the skeleton has highlighted the intrusive nature of some of the material in this barrow and dates at least one of the burials to 900 to 800 BC. This issue is being recognised as more dates are obtained on material which might not have been considered important in the past due to assumptions about site chronology, and also as re-deposited or intrusive remains are investigated (McKinley, Sept. 2013, pers. comm.).

There are many other ways in which the data can be interrogated. One of the most important when looking for the 'invisible' skeletal material may be the consideration of burial monuments themselves, particularly barrows. These are known to be present in relatively large numbers for much of later prehistory in one form or another, but, as discussed above, they tend to disappear along with the skeletal remains towards the end of the Bronze Age (Grinsell, 1990: 47; Darvill, 2010: 221). This we already know, but there are many records of *probable* and *possible* barrows which, alongside those which are unexcavated, might be used in conjunction with the presence of skeletal remains to start to estimate the numbers of burials which may have taken place for those periods. Associating skeletal remains with certain barrow types where we have them, and then extrapolating these numbers to those which we can record from sources such as aerial photography as *probable* or *possible* burial monuments, will allow better estimations of how many burials may actually be 'invisible' in terms of skeletal remains. If we can then formulate an approximate minimum value for the numbers of 'invisible dead' we might be dealing with, we can perhaps start to estimate how many individuals we might be missing from our current chronological reconstructions and spatial burial plots.

At present, age and sex information has been recorded for only some of the sites from which the data are available, but an example of another way in which we might look at the 'invisible' part of the population might relate to a presentation of the ages and, perhaps, look for neonates in various periods. When the database in its current form is used to plot out neonates as opposed to all aged individuals, there is a definite preponderance of data in the south of England. The data are not currently representative, but it may be appropriate to look for the 'invisible' neonates in other areas and to consider

whether, when the data are plotted onto geological base maps, the burial conditions have affected preservation to such an extent that identification of such remains is more restricted than other remains (Brothwell, 1972). Another scenario to consider may be to check whether neonates are proportionally more prevalent amongst the Roman sites, which may affect their geographical spread or indicate the practice of infanticide (e.g., Mays and Eyers, 2011), or say something about the treatment of neonates in the earlier periods and the burial rites to which they might be entitled at varying points in the timeline (e.g., Waterman and Thomas, 2011).

4. Conclusions

In order to view the big picture of corpse disposal, burial, and skeletal remains in the archaeological record, we need a detailed and extensive database which brings together the data we already have in a vast quantity of reports, both published and unpublished, alongside existing databases which are fragmentary and supply only parts of the data we need. The examples of spatial and chronological analyses discussed in this paper give a flavour of what may become possible as the database expands.

This is an initial attempt to bring these data together and to start using GIS to analyse patterns, both in space and through time. Comparing the data for the numbers of individuals found in the archaeological record with demographic estimates or proxies (e.g., by using pollen records to investigate human presence fluctuations over time), we can then be clearer about when, where and why the dead appear to be 'invisible'. This will make it easier to consider whether this 'invisibility' is related to corpse disposal methods themselves or to biases in the archaeological or excavation record.

The hope for the future is that this is only the first phase of an exciting process of building a database we can all use to investigate a range of research issues. The flexibility of the database has been designed with that expansion and long-term view in mind.

Acknowledgements

The project is being undertaken at Durham University with funding provided by the John Templeton Foundation. The database template and data entry system adapted for this project was originally designed by the Vanishing Landscape and Fragile Crescent Project team (https://www.dur.ac.uk/fragile_crescent_project/project_members/). In particular we should mention Dr Rob Dunford who was largely responsible for designing the infrastructure of the original database. Jacqueline McKinley at Wessex Archaeology is thanked for her input into the discussion relating to intrusive and re-deposited remains which start to fill in the gaps in the record. Some of the data mentioned in the text is from the AHRC-funded Beaker People Project (PIs: Mike Parker Pearson, Mike Richards and Andrew Chamberlain).

Note

1 This paper was originally written in 2013. See Jay and Scarre 2017 for a more up to date view of progress made on the database.

References

Allen, M.J., Gardiner, J. 2009. If you go down to the woods today; a re-evaluation of the chalkland postglacial woodland; implications for prehistoric communities, in: Allen, M.J., Sharples, N.M., O'Connor, T. (Eds.), *Land and People: Papers in Memory of John G. Evans*. Oxbow Books, Oxford, pp. 49–66.

Barber, B., Bowsher, D. 2000. *The Eastern Cemetery of Roman London: Excavations 1983–1990*. Museum of London Archaeology Service (MoLAS), London.

Bradbury, J., Davies, D., Jay, M., Philip, G., Roberts, C., Scarre, C. 2016. Making the dead visible: Problems and solutions for the "big" picture approaches to the past, and dealing with large "mortuary" datasets. *Journal of Archaeological Method and Theory* 23, 561–591.

Bristow, P.H.W. 2001. Behaviour and belief in mortuary ritual: Attitudes to the disposal of the dead in southern Britain 3500bc-AD43. *Internet Archaeology* 11, https://doi.org/10.11141/ia.11.1 (accessed 22nd May, 2018).

Brothwell, D.R. 1972. Palaeodemography and earlier British populations. *World Archaeology* 4(1), 75–87.

Brothwell, D.R. 1981. *Digging Up Bones*. Cornell University Press, New York.

Brück, J. 1995. A place for the dead: The role of human remains in Late Bronze Age Britain. *Proceedings of the Prehistoric Society* 61, 245–277.

Carter, S., Hunter, F., Smith, A. 2010. A 5th century BC Iron Age chariot burial from Newbridge, Edinburgh. *Proceedings of the Prehistoric Society* 76, 31–74.

Champion, T., 2009. The later Bronze Age, in: Hunter, J., Ralston, I. (Eds.), *The Archaeology of Britain: An Introduction from Earliest Times to the Twenty-First Century*. Routledge, London, pp. 126–148.

Darvill, T. 2004. *Long Barrows of the Cotswolds and Surrounding Areas*. Stroud, Tempus.

Darvill, T. 2010. *Prehistoric Britain*. Routledge, Abingdon.

Dobbie, K.E., Bruneau, P.M. C., Towers, W.E. (Eds.), 2011. *The State of Scotland's Soil*. http://www.soils-scotland.gov.uk/documents/15130508_SOSreport.pdf, Natural Scotland (accessed 1st September, 2015).

Dyer, J.F. 1963. Earthworks of the Icknield Way: Part one. *Bedfordshire Magazine* 8, 161–166.

English Heritage, Pastscape. http://www.pastscape.org.uk/textpage.aspx (accessed 1st September, 2015)

Green, M. 2000. *A Landscape Revealed: 10,000 years on a Chalkland Farm*. Tempus, Stroud.

Grinsell, L.V. 1990. *Barrows in England and Wales*. Shire Publications, Princes Risborough.

Harrison, S. 2003. The Icknield Way: Some queries. *Archaeological Journal* 160, 1–22.

Jay, M., Parker Pearson, M., Richards, M.P., Nehlich, O., Montgomery, J., Chamberlain, A., Sheridan, A. 2012. The Beaker People Project: An interim report on the progress of the isotopic analysis of the organic skeletal material, in: Allen, M.J., Sheridan, A., McOmish, D. (Eds.), *The British Chalcolithic: People, Place and Polity in the Later 3rd Millennium*, Prehistoric Society Research Paper No. 4. The Prehistoric Society and Oxbow Books, Oxford, pp. 226–236.

Jay, M., Scarre, C. 2017. Tracking the dead in the Neolithic: The 'Invisible Dead' in Britain, in: Bradbury, J., Scarre, C. (Eds.), *Engaging with the Dead: Exploring changing human beliefs about death, mortality and the human body*. Oxbow Books, Oxford, pp. 7–23.

Lanting, J.N., Brindley, A.L. 1998. Dating cremated bone: The dawn of a new era. *The Journal of Irish Archaeology* 11, 1–7.

Lawrence, D., Bradbury, J., Dunford, R. 2012. Chronology, uncertainty and GIS: A methodology for characterising and understanding landscapes of the Ancient Near East. *eTopoi Journal for Ancient Studies* 3, 353–359.

Manby, T.G. 1980. The Yorkshire Wolds: field monuments and arable farming, in: Hinchliffe, J., Schadla-Hall, R.T. (Eds.), *The Past under the Plough*. Department of the Environment, Directorate of Ancient Monuments and Historic Buildings, Occasional Paper No. 3, London, pp. 60–68.

Mays, S., Eyers, J. 2011. Perinatal infant death at the Roman villa site at Hambleden, Buckinghamshire, England. *Journal of Archaeological Science* 38, 1931–1938.

Mortimer, J.R. 1905. *Forty Years' Researches in British and Saxon Burial Mounds of East Yorkshire*. A. Brown & Sons, London.

Natural England, 2002. *GIS Digital Boundary Datasets, Ancient Woodland Inventory (Provisional) for England - Digital Boundaries*. http://www.gis.naturalengland.org.uk/pubs/gis/tech_aw.htm (accessed 20th August, 2013)

Nielsen-Marsh, C.M., Smith, C.I., Jans, M.M.E., Nord, A., Kars, H., Collins, M.J. 2007. Bone diagenesis in the European Holocene II: Taphonomic and environmental considerations. *Journal of Archaeological Science* 34, 1523–1531.

Parker Pearson, M., Chamberlain, A., Jay, M., Richards, M.P., Sheridan, A. (Eds.), In prep. *The Beaker People: Isotopes, Mobility and Diet in Prehistoric Britain*. Prehistoric Society Monograph, Oxbow Books, Oxford.

Ritchie, J.N.G. 1978. The Stones of Stenness, Orkney. *Proceedings of the Society of Antiquaries of Scotland* 107, 1–60.

Stead, I.M. 1976. La Tène burials between Burton Fleming and Rudston, North Humberside. *Antiquaries Journal* 56, 217–226.

Waterman, A.J., Thomas, J.T. 2011. When the bough breaks: Childhood mortality and burial practice in Late Neolithic Atlantic Europe. *Oxford Journal of Archaeology* 30, 165–183.

Whimster, R. 1981. *Burial Practices in Iron Age Britain*. BAR British Series 90, Oxford.

Zazzo, A., Saliège, J.-F., Lebon, M., Lepetz, S., Moreau, C. 2012. Radiocarbon dating of calcined bones: Insights from combustion experiments under natural conditions. *Radiocarbon* 54, 855–866.

Diverse Burial Practices from a Late Medieval Cemetery in Somerset

Heidi Dawson-Hobbis

It is often assumed that in the late medieval period burial practice was highly standardised. We expect to find individuals who were simply shrouded, laid out supine in separate graves, on a west-east alignment with their heads to the west, and with no evidence for grave inclusions. While this may be seen as the norm, recent research into medieval burial practice has shown that funerary contexts in this period can include a diverse range of rites. This paper will present some of the evidence for diverse burial practice from the excavation of part of the church and lay cemetery of the priory of SS Peter and Paul, Taunton, Somerset, which was excavated in 2005. Interesting practices include the use of ash and charred planks as grave inclusions, the reuse of stone-lined graves within the nave, perhaps by family members, the possible transfer of exhumed remains from one site to another, and the use of graves for more than one individual.

Keywords Disarticulation; Translated; Ash; Charcoal; Church; Cemetery

1. Introduction

In general, it is assumed that late medieval (AD 1066–1540) Christian burial practice tends to be fairly homogenous. The body is almost always laid in a supine position on a west-east alignment and graves tend to be simple and without the inclusion of grave goods. However, although often subtle, there are variations in burial practice that have been observed from archaeological excavations of cemeteries from the period. There was a tendency in the medieval period to classify individuals according to their status whether defined by religious role, lineage, gender, age, or even pathology, as in the case of people with leprosy (Gilchrist and Sloane, 2005). These classifications may, therefore, also appear in death through the location of burial, the positioning of the body, and the presence of inclusions within the grave. Variation in the positioning of the body can include deviations from the norm for alignment, placement of the body on its side or front (prone), and differential positioning of the limbs. Inclusions within the grave can be; burial within a coffin, the presence of grave goods or linings, and the use of stone structures within the grave. The re-use over time of defined burial areas caused the regular displacement,

disarticulation and dispersal of individuals buried within late medieval cemeteries. The disturbed remains were sometimes collected and buried in charnel pits, as seen at St Helen-on-the-Walls (Dawes and Magilton, 1980), or merely re-interred as backfill to later graves, as was often the case at the site that is the focus for this paper, the church and cemetery of the priory of SS Peter and Paul, Taunton.

Excavations at Priory Avenue, Taunton were undertaken in 2005 by Context One Archaeological Services in advance of development on the site. The excavation area was known to have been located within the cemetery of the priory, and the west end of the priory church, whose exact whereabouts had been previously unknown, was also uncovered. A range of diverse burial practices have been revealed by these excavations, which will be discussed in detail here. The land at Taunton was granted to the priory in AD 1158 by Henry of Blois, Bishop of Winchester (Bush, 1984). The historical records suggest that the land appears to have been developed slowly, with no mention of the priory church for over a century until a charter in AD 1249 (Page, 1911). Documentary evidence implies that the priory was an extremely wealthy house and there is evidence of repeated building and rebuilding (Gathercole, 2002). In AD 1277 there are references indicating that the priors were having difficulty financing the building of the church in 'a style of great magnificence' (Hugo, 1860: 11), with the completion date being as late as AD 1342 (Bush, 1984). There is a mention of the cemetery in AD 1349 (Hugo, 1860), but it is likely that the priory cemetery was in use before the completion of this church[1]. Evidence from the excavations in 2005 suggests that there may have been an earlier church on the site. One excavated burial (skeleton 5017) had been cut by the foundations of the church indicating the presence of a cemetery predating the excavated church building. Broken fragments of at least three grave slabs had been used for the foundations of the pier bases which ran along the nave of the church; these have been dated typologically to the early 13th to early 14th century AD and indicate reuse of material from an earlier church (Place, 2009).

The cemetery at the site of the new priory at Taunton may therefore have been in use from the middle of the 12th century and would have been the burial place for all the people of Taunton, as well as the inhabitants of the surrounding area, until AD 1446, when St Mary's cemetery was consecrated (Bush, 1977). The Somerset Wills show that burial in the priory was still popular after this date (Weaver, 1901, 1903, 1905), and continued until the priory went out of use as a religious house on the 12th February 1539 (Hugo, 1860).

Burial evidence from Taunton comes from both intra and extra mural burial practice. Location of burial has been cited as an important indicator of status with more and less holy areas defined within the church and cemetery (Gilchrist and Sloane, 2005). Daniell (1997) describes the late medieval church as a series of concentric rings with the most holy area being the high altar at the east end of the church, the holiness lessening to the west end and out into the cemetery. All the consecrated ground would have been enclosed within the boundaries of the cemetery. Intramural burials (within the church buildings) would have been reserved for those of the religious orders (usually closest to the east end), or to those able to afford such a privilege, with lay burial more usual away from the high altar (Daniell, 1997). At Taunton only the west end of the nave was uncovered, the east end (if still surviving) being located beneath roads and housing from the Victorian era. It is likely, therefore, that the individuals buried here are members of the lay community and the presence of interments of males, females and children within the church confirms this. Daniell (1997) also suggests that the cemetery itself was divided into areas with some being more desirable than others. His conclusion was that the south side was more favoured than the north, and that the churchyard cross was probably a popular site for burial.

The majority rite for this period, and that seen within the cemetery at Taunton, would have been burial of the body wrapped in a shroud and laid within an earth-cut grave. Evidence for coffined burial is less frequently uncovered, and burial within stone lined graves and tombs was undoubtedly reserved for those of high status. Thirteen out of 97 adults and two out of 93 sub-adults provided good evidence for burial within a coffin at Taunton, consisting of nails surrounding the body or soil staining indicating decayed wood. The fear of disease may also have led to coffined burial; for example, a high number of coffins were in evidence at the Black Death Cemetery at East Smithfield, London (Grainger *et al.*, 2008). For most people, although they may have been carried to church, and to the grave, within a coffin, they would have been laid in the ground merely wrapped in a shroud (Daniell, 1997), the coffin being reused for another funeral.

2.   Burials within the Church

Four stone-lined graves were uncovered at Taunton running along the nave of the church, of which three were excavated; all showed signs of disturbance. Williams (2003) discusses the importance of memory within the period and how both the memory of individuals and society as a whole can be manipulated through ritual acts involving the placement of the body, artefacts, and monuments, determining who or what shall be remembered. The disturbance of these graves could be interpreted as a form of remembrance of the dead but equally as a means of forgetting the dead as individuals. Grave 830 contained a vast amount of disarticulated remains, which had been placed on top of the complete articulated inhumation at the base of the stone-lined grave. The disarticulated remains included those of at least three adults, one female and two males, and at least two children. The older child aged to around 10–12 years and the younger less than 2 years of age. The articulated individual (skeleton 851) is a mature male with evidence for diffuse idiopathic skeletal hyperostosis on the spine (Dawson and Robson Brown, 2009), a pathology associated with high status individuals at other medieval sites (Rogers and Waldron, 2001). It can only be speculated whether the disarticulated remains come from the disturbed previous occupants of this stone-lined grave, or were merely disturbed when the grave was cut into the floor of the nave, with the bone elements being placed into the grave after the burial of skeleton 851. Figure 4.1 depicts skeleton 851 within grave 830 with the disarticulated bones laid in a regular way appearing to create cross motifs. As many cemeteries were in use for long periods of time in Britain, the disturbance of previous burials was common. Often the disturbed remains were included in the backfill of the newly cut grave, but occasionally they were placed in a deliberate and symbolic way. For example, a cross, made from long bones, was placed over a child at St Anne's, Coventry, Warwickshire (Gilchrist and Sloane, 2005), along with the placement of a skull into a 'skull and crossbones' motif, and at Whithorn, Scotland an infant was placed over the remains

Figure 4.1 Skeleton 851 from stone-lined grave 830. Showing the placement of the disarticulated remains with the long bones laid in a cross type motif (© Context One Archaeological Services).

of an adult woman, her disturbed arm bones being used to form a cross (Cardy, 1997).

A second stone-lined grave (833) had been cut through at both ends to allow for the later insertion of two more individuals. One of these was an elderly male with healed fractures to both lower legs and osteoarthritis of the spine, wrist and hands (Dawson and Robson Brown, 2009); the other was a middle-aged male. These graves respected each other but whilst the alignment of these burials was correct they had not respected the earlier grave, cutting through both the east and west ends of it. All that remained of the original occupant were portions of the left and right femora, left tibia, and left fibula. These findings suggest that the location of these stone-lined graves may have been retained through memory, or the use of grave markers, as the resting places of certain families, and that the disarticulated remains contained within them may have been members of the same family. Being buried close to family members was important to individuals of the period as some of the wills relating to the priory of SS Peter and Paul, Taunton attest. Alison Togwell requested burial 'in (the) west end beside the body of my husband' (Weaver, 1903), John Tose 'in the north part of the said churche in the place joined and annexed unto the Tombe where the bodies of my fadre (father) Alexander Tose and my modre

(mother) Jone Tose restith' (*ibid.*), whilst Alexander Tuse makes the request to be buried 'in the same monument in which Joan, my former wife was buried' (Weaver, 1901).

Whilst within these stone-lined graves, and in the cemetery as a whole, the constant disturbance of earlier burials was normal, there was one interesting 'burial' within the nave which may consist of a translated burial. This 'bone bundle' (Figure 4.2) contained the disarticulated remains of at least two adult individuals. These remains appear to have been kept in association to some degree and re-interred as a wrapped bundle, rather than randomly backfilled, as the majority of disturbed individuals would have been. The remains are mostly from an adult male of less than 50 years of age, with a few elements appearing to be from an adult female, including a left temporal bone. The inclusion of bones appearing to be from a female individual implies that these elements are more likely to have come from a lay burial than from that of one of the religious community, but also that this person was, or people were, remembered and viewed as important enough to be reburied individually rather than as part of the disturbed collective. An example of a much more complete translated burial of the period, is known from the priory of St Oswald, Gloucester. The complete skeleton of this individual appears to have been recovered,

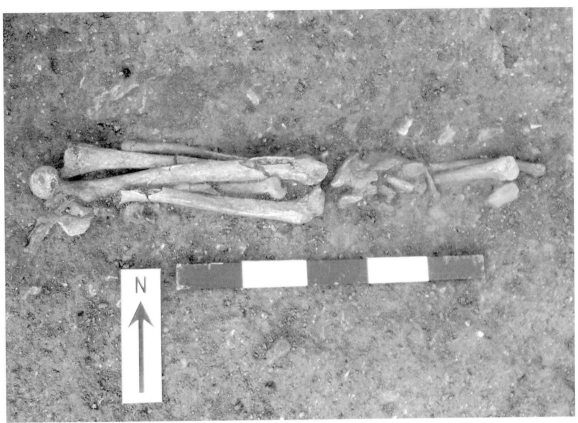

Figure 4.2 'Bone bundle'. A collection of bones excavated from within the nave which appear to have been wrapped together into a bundle (© Context One Archaeological Services).

and the presence of nails indicates that the remains were buried within a box or chest (Highway and Bryant, 1999). No nails were recovered in association with the 'bone bundle' at Taunton.

3. Ash and Charred Plank Burials

Two examples of a fairly rare rite are present at Taunton, that of laying the deceased on a bed of ash which had been placed within a coffin. One such burial was located in the nave of the church (skeleton 908) and contained the remains of a young adult male. Another burial (skeleton 1642), of a probable middle-aged adult female, was excavated from within the cemetery area to the west of the church. The ash from both burials has been analysed and contained fine ash, charcoal, burnt and unburnt fish bones, and charred cereal grains. Wheat was present with skeleton 908 and barley with skeleton 1642 (Vaughan-Williams, 2009). Wheat was seen as a higher status grain in the period (Adamson, 2004), and it is notable that it was found in the intramural burial, whereas barley was recovered from the cemetery burial. Shellfish remains, possibly of edible crab, and burnt sheep remains were also present within the ash layer associated with skeleton 908 (Higbee, 2009). The charcoal associated with skeleton 908 consisted mainly of alder (*Alnus*) with some fragments of ash (*Fraxinus*); that associated with skeleton 1642 was less well preserved but fragments of oak (*Quercus*), alder, and beech (*Fagus*) were identified (Vaughan-Williams, 2009).

A silver halfpenny from the reign of Edward III, dated to AD 1344–52, was also associated with the ash burial of skeleton 1642. This burial had been badly disturbed by inter-cutting of later burials but the coin was close to where the skull would have been, and it may be speculated that it was some form of symbolic offering.

Various reasons for the inclusion of ash and charcoal in these burials have been suggested. Functional explanations for ash burials focus on the ash being used to absorb bodily fluids (Hadley, 2001), perhaps indicating a delay in burial. Some notable movement was seen of the skull and part of the pelvis of skeleton 908 (Figure 4.3). This may support the association with absorbency needed for delayed burial due to the disarticulation of elements, such as detaching of the skull, possibly occurring due to movement of the corpse when putrefaction is already well established. Another example, which may support this theory, is an ash burial with partial disarticulation from the East Smithfield cemetery (Grainger *et al.*, 2008). It may also have been believed during the period that the charcoal or ash would prevent further decay as well as soaking up odours and fluids (Gordon 2014). In general, due to the extra effort involved, most authors see ash burials as a high status rite (Heighway and Bryant, 1999; Daniell, 1997), and whilst both Taunton examples may support this interpretation, with one burial within the church and the other associated with a coin, the examples from East Smithfield do not. Gilchrist (2008) has suggested that these ashes represent the rakings of domestic hearths and

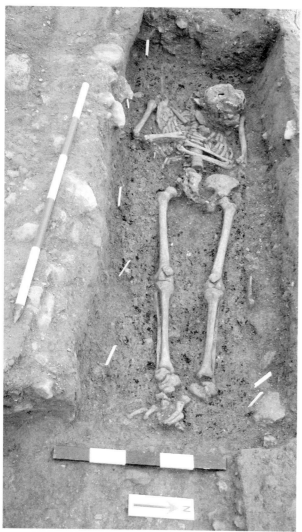

Figure 4.3 Skeleton 908. The ash burial from within the nave of the church, note the movement of the skull (© Context One Archaeological Services).

may have been involved in rituals to protect the living. She cites an earlier rite of burning grain to protect the house from the dead and to stop them returning. The high occurrence of ash burials in the East Smithfield Black Death cemetery is also indicative of a symbolic use associated with disease (Grainger *et al.*, 2008). It could also be postulated that these burials may suggest a form of remembrance in the raking of the hearth symbolising the last days and meals of the deceased, but equally as a way of forgetting the dead and moving on, clearing out the hearth and sending the rakings off with the dead to leave a new hearth for the living.

Two burials involving charred planks were also excavated at Taunton. Skeleton 342 from the area to the west of the church represents an unexcavated adult; only the left foot and lower leg were exposed. This individual had been laid on a charred plank, which was probably part of a coffin base; a single iron nail was also noted in-situ suggesting this formed the base of a coffin as opposed to a single plank. One other example of a charred plank comes from a burial located beyond the extent of the

supposed boundary ditch of the cemetery; three adult males were excavated from this area. Burial outside the cemetery area is rare and often cited as a place for unbaptised infants, the excommunicated, and suicides (Orme, 2001). Skeleton 2203, a middle-aged male, had a charred plank laid over their body. The surrounding soil, above the body, showed a distinctive darker stain indicating the presence of a coffin lid, with only one of the planks having been charred; there were, however, no coffin nails present. Daniell (1997) suggests that fewer than six nails would have been insufficient to hold a coffin together, whilst Rodwell (1989) suggests at least a dozen would be necessary. However, some coffins may have been held together with wooden dowels or pegs, such as the examples from Barton-upon-Humber, including the coffin of a child (Waldron, 2007), or a combination of wood and iron pegs may have been used. Simple wooden boards rather than coffins may also have been used beneath or above the body; therefore an absence of nails does not necessarily equate to the absence of a coffin. The purpose of these charred planks is hard to determine; the single plank over skeleton 2203 may have acted as a marker to avoid disturbance of this burial which was located on the very outskirts of the cemetery, beyond what has been interpreted as the cemetery enclosure ditch. This may have been due to fear of the deceased, possibly related to their cause of death, or for some reason associated with the individual in life. It could be suggested that the example beneath skeleton 342 was a high status coffin burial; alternatively the incorporation of a charred base could have been due to a reused burnt piece of wood being a cheaper option. The plank laid over skeleton 2203 may also have been a reused piece of wood for a coffin, but its location over the body is suggestive of its use as a marker, to avoid later disturbance to this grave. Reasons could be due to a sudden or unexplained death, or as a means of warning against some infectious disease such as the Black Death, which would have affected the inhabitants of Taunton in the 14th century as it did throughout the country. The liminal situation also suggests exclusion of this burial.

4. Double Burials

Another rite seen at Taunton is that of double burial. Double burials involve two individuals being buried within the same grave at the same time. These are often mothers and newborn children who both died at or soon after birth. There were several examples excavated at St Helen-on-the-Walls of females buried in close association with infants (Dawes and Magilton, 1980), and examples of older children buried with adults are also known (Stroud and Kemp, 1993).

The three double burials recovered during the excavations at Taunton in 2005 all involve children. In the area at the northern limit of the cemetery, skeleton 2080 (aged 8–10 years) and skeleton 2089 (of a similar age) were laid next to each other on the natural bedrock with their hands touching (Figure 4.4). To the area west of

Figure 4.4 Double burial of skeletons 2080 and 2089. These two sub-adults appear to have been placed in the same grave, note the entwined hand bones (© Context One Archaeological Services).

the church skeleton 1617 (aged 9–11 years) and skeleton 1620 (aged 4–6 years) were also laid together, with hand bones entwined. The third example, skeleton 1205 (aged 6–8 years) and skeleton 1206 (aged 7–9 years), does not involve the laying over of their hands, so it is less certain that the two individuals were buried together. However, this appears likely due to their close positioning in the grave and that there is no evidence for disturbance, which would have been the case if one of the burials had occurred at a later date. In all cases the children had been laid supine with their skulls facing south, with the exception of one skull, which had been disturbed in antiquity. An example of another double burial involving two similar age children comes from the excavations of the priory of St Gregory, Canterbury (Hick and Hicks, 2001).

The placing of two children together in a single grave may be for monetary reasons, involving less cost. Emotional reasons should also be considered; if two children in the community, possibly friends or relations, died around the same time it may have provided comfort to their parents if they were buried together. They could also represent siblings (or even possibly twins in the case of skeleton 2080 and 2089) who had both died of the same disease; this theory could be resolved in the future by ensuring such burials were excavated under forensic conditions to allow for testing, without the fear of modern contamination, for any surviving DNA in the bones.

5.   Conclusion

The excavation of this one site demonstrates that the funerary archaeology of the late medieval period can involve diverse practices, the reasons for which are not always easy to explain. The disturbance of individuals associated with the stone-lined graves within the nave of the church may suggest some form of remembrance of the dead, as does the possible translated burial. However, forgetting the dead also appears to be a theme, with bodies becoming disarticulated and scattered as part of a cemetery collective. For the ash burials, functional reasons to do with decay of the body have been discussed, but these may also have a superstitious or ritual function, as may the charred planks associated with two of the burials. Whilst the charred plank over skeleton 2203 may be interpreted as some form of marker, the presence of a charred plank beneath the body of skeleton 342 is harder to interpret. It is unlikely to be a marker and therefore may have other symbolic associations, or be the functional recycling of a piece of wood for the coffin base. Further research into these interesting burial practices from sites across the UK and Europe is needed to determine the likelihood of the reasons given for such burial inclusions.

Acknowledgements

Acknowledgement goes to Professor Kate Robson Brown, the author's PhD supervisor, mentor and friend, Mr Richard McConnell Director of COAS, and all the team at COAS who worked on the excavation of the church and cemetery of SS Peter and Paul, Taunton.

Note

1   It is unlikely burials would have continued at the original site of the priory, which was moved due to the building of Taunton castle.

References

Adamson, M.W. 2004. *Food in Medieval Times*. Greenwood Press, Westport.

Bush, R.J.E. 1984. Section IV The Priory: The Priory of St Peter and St Paul, in: Leach, P. (Ed.), *The Archaeology of Taunton Part 1: Excavation and Fieldwork*, Western Archaeological Trust Excavation Monograph Number 8. Sutton Publishing, Gloucester, pp. 104–106.

Bush, R. 1977. *The Book of Taunton*. Barracuda Books Limited, Chesham.

Cardy, A. 1997. The human bones, in: Hill, P. (Ed.), *Whithorn and St Ninian: The Excavation of a Monastic Town 1984–91*, The Whithorn Trust. Sutton Publishing, Gloucester, pp. 519–592.

Daniell, C. 1997. *Death and Burial in Medieval England 1066–1550*. Routledge, London.

Dawes, J.D., Magilton, J.R. 1980. *The Cemetery of St Helen-on-the-Walls, Aldwark: The Archaeology of York, The Medieval Cemeteries*. York Archaeological Trust/CBA, York.

Dawson, H., Robson Brown, K. 2009. The human remains from the medieval church and cemetery of the priory of St Peter and St Paul, Priory Avenue, Taunton, in: Place, C. (Ed.), *Former County Garage, Priory Avenue, Taunton, Somerset: An Archaeological Excavation*. Context One Archaeological Services (unpublished excavation report).

Gathercole, C. 2002. *An Archaeological Assessment of Taunton*. English Heritage Extensive Urban Survey, Somerset County Council, Taunton.

Gilchrist, R. 2008. Magic for the dead? The archaeology of magic in later medieval burials. *Medieval Archaeology* 52, 119–159.

Gilchrist, R., Sloane, B. 2005. *Requiem: The Medieval Monastic Cemetery in Britain*. Museum of London Archaeological Service, London.

Gordon, S. 2014. Disease, Sin and the Walking Dead in Medieval England, c. 1100–1350, in: Gemi-Iordanou, E., Gordon, S., Matthew, R., McInnes, E., Pettitt, R. (Eds.), *Medicine, Healing and Performance*. Oxbow Books, Oxford, pp. 55–70.

Grainger, I., Hawkes, D., Cowal, L., Mikulski, R. 2008. *The Black Death Cemetery, East Smithfield, London*. MOLAS Monograph 43, Museum of London Archaeology Service, London.

Hadley, D.M. 2001. *Death in Medieval England*. Tempus, Stroud.

Heighway, C., Bryant, R. 1999. *The Golden Minster: The Anglo-Saxon Minster and Later Medieval Priory of St Oswald at Gloucester*. CBA Research Report 117, Council for British Archaeology, York.

Hicks, M., Hicks, A. 2001. *St Gregory's Priory, Northgate, Canterbury Excavations, 1988-1991*. Canterbury Archaeological Trust, Canterbury.

Higbee, L. 2009. Animal bone, in: Place, C. (Ed.), *Former County Garage, Priory Avenue, Taunton, Somerset: An Archaeological Excavation*. Context One Archaeological Services (unpublished excavation report).

Hugo, Rev. T. 1860. Taunton Priory. *Somerset Archaeological and Natural History Society Proceedings during the year 1859*. Volume IX, 1–127.

Orme, N. 2001. *Medieval Children*. Yale University Press, New Haven.

Page, W. 1911. *A History of the County of Somerset Volume 2*, Victoria County History online at http://www.british-history.ac.uk/source.aspx?compid=40934 (accessed 9th December, 2010).

Place, C. 2009. *Former County Garage, Priory Avenue, Taunton, Somerset: An Archaeological Excavation*. Context One Archaeological Services (unpublished excavation report).

Rodwell, W. 1989. *Church Archaeology*. Batsford/English Heritage, London.

Rogers, J., Waldron, T. 2001. DISH and the monastic way of life. *International Journal of Osteoarchaeology* 11, 357–365.

Stroud, G., Kemp, R.L. 1993. *Cemeteries of the Church and Priory of St Andrew Fishergate, The Archaeology of York: The Medieval cemeteries*. York Archaeological Trust/CBA, York.

Vaughan-Williams, A. 2009. Archaeobotanical assessment, in: Place, C. (Ed.), *Former County Garage, Priory Avenue, Taunton, Somerset: An Archaeological Excavation*. Context One Archaeological Services (unpublished excavation report).

Waldron, T. 2007. *St Peter's Barton-upon-Humber, Lincolnshire: A Parish Church and its Community. Volume 2 The Human Remains*. Oxbow Books, Oxford.

Weaver, Rev. F.W. 1901. *Somerset Medieval Wills 1383–1500*. Somerset Record Society 16.

Weaver, Rev. F.W. 1903. *Somerset Medieval Wills 1501–1530*. Somerset Record Society 19.

Weaver, Rev. F.W. 1905. *Somerset Medieval Wills 1531–1558*. Somerset Record Society 21.

Williams, H. 2003. Remembering and forgetting the medieval dead, in: Williams, H. (Ed.), *Archaeologies of Remembrance: Death and Memory in Past Societies*. Kluwer Academic/Plenum Publishers, New York, pp. 227–254.

Whole Genome Analysis of *Mycobacterium tuberculosis* in 18th-Century Natural Mummies from Vác, Hungary

Helen D. Donoghue, Mark Spigelman, Ildikó Pap, Ildikó Szikossy, Oona Y.-C. Lee, David E. Minnikin, Gurdyal S. Besra, Andrew Millard, Martin J. Sergeant, Jacqueline Z.-M. Chan and Mark J. Pallen

The large collection of skeletalised and partially mummified human remains from the 18th-century Dominican Church of Vác, Hungary, has facilitated the study of tuberculosis in a wealthy population dating from the pre-industrialised and pre-antibiotic eras. The church and civic archives enabled the identification of some family groups and provide a context to support the study of the tuberculosis that was endemic in this population. Conventional ancient DNA (aDNA) analysis using Polymerase Chain Reaction (PCR) demonstrated ten years ago that there were two distinct genotypes of Mycobacterium tuberculosis *in the population and three members of one family were each infected with a different strain of the organism. This scenario is consistent with communal spread rather than transmission within the family. The application of metagenomics – open-ended sequencing of DNA recovered directly from samples, without target-specific amplification or enrichment – provides global coverage of the genome. It also enables minority genetic sequences to be detected and assessed and finds strain-specific sequences absent from any other known genomes. Using metagenomics we found evidence of co-infection of an individual with* Mycobacterium tuberculosis *of two distinct genotypes. These fall within the cluster of the MTB Haarlem European lineage and resemble strains currently found in Germany.*

Keywords Ancient DNA; Lipid analysis; Metagenomics; Mixed infections; Tuberculosis

1. Introduction

1.1. Emergence of Palaeomicrobiology To Study Ancient Tuberculosis

Tuberculosis is an ancient human disease, recognised by distinct skeletal changes. The most characteristic condition is kyphosis – the collapse of the spine and fusion of vertebrae known as Pott's disease. This has been found in human remains from pre-dynastic Egypt and was described by classical authors such as Hippocrates and Galen. Mummified remains enable imaging and histological examination, which led to the first identification of a case of tuberculosis in pre-Columbian South America from AD 700 (Allison *et al.*, 1975). The development of DNA amplification by the polymerase chain reaction (PCR) enabled the detection of *M. tuberculosis* aDNA in ancient skeletons (Spigelman and Lemma, 1993) and mummified human remains (Salo *et al.*, 1994), resulting in the establishment of the new research field of palaeomicrobiology (Donoghue,

2008). These early studies verified conclusions reached from palaeopathology and also answered the historical questions of whether there was tuberculosis in the Far East (Spigelman and Lemma, 1993) and Peru (Salo *et al.*, 1994), before any known European contact. In both cases the answer was in the affirmative. Subsequent work established that some bones and tissues without typical pathology, or even no pathology at all, contain tuberculosis biomarkers (Zink *et al.*, 2003), leading to criticisms of the methodology (Donoghue and Spigelman, 2006; Willerslev and Cooper, 2006). However, it is now realised by palaeopathologists that such findings are consistent with the recent historical record from the pre-antibiotic era, where skeletal involvement occurred only in around 3–5% of tuberculosis cases. However, a useful practice was introduced, of using independent collaborators to confirm aDNA findings, initially by replicating PCR studies (Spigelman *et al.*, 2002; Taylor *et al.*, 2009). Other non-DNA techniques are now being used increasingly for both detection and independent corroboration of ancient biomolecules (Tran *et al.*, 2011), including proteomics (Corthals *et al.*, 2012) and imaging (Holloway *et al.*, 2013). The direct detection of mycobacterial cell wall lipid biomarkers has been especially valuable, as these have sufficient specificity to identify the *M. tuberculosis* complex (MTBC) that contains the pathogenic species, yet can be more resistant to diagenesis than mycobacterial DNA (Donoghue *et al.*, 1998; Lee *et al.*, 2012; Minnikin *et al.*, 2012).

1.2.   The nature of the *M. tuberculosis* genome past and present

Sequencing of the total *M. tuberculosis* genome revealed that members of the MTBC exhibit sequential deletions (Brosch *et al.*, 2002). With the exception of the *Mycobacterium canettii* strains, found in the Horn of Africa and with many ancestral features, members of the MTBC appear to have co-evolved with their host after undergoing an evolutionary bottleneck (Gutierrez *et al.*, 2005; Smith *et al.*, 2009) possibly associated with the microbe adopting an intracellular lifestyle. Comparison of the *M. tuberculosis* and *Mycobacterium bovis* genomes demonstrated that *M. tuberculosis* was more ancestral and therefore could not have originated from modern *M. bovis* (Hershkovitz *et al.*, 2015). The extremely limited horizontal gene transfer within the post-bottleneck MTBC enables evolutionary changes to be monitored by deletion analysis. Different *M. tuberculosis* lineages are associated with different human populations around the globe even today (Gagneux *et al.*, 2006). The emergence of human infectious diseases appears to be associated with human population density (Blaser and Kirschner, 2007), becoming apparent in the early Neolithic period. In the case of tuberculosis, this association is continuing with the emergence of highly transmissible and virulent strains of *M. tuberculosis* in major cities that have a long record of continuous habitation (Barnes *et al.*, 2010). Whole genome analysis has now been used to analyse 259 modern MTBC strains to characterise global diversity and to reconstruct the evolutionary history of *M. tuberculosis* (Comas *et al.*, 2013). Modelling suggested the emergence of tuberculosis in Africa about 70,000 years ago. Genomic analysis of 4995 human mitochondria indicated a parallel evolutionary scenario for anatomically modern humans, but no direct link between the two evolutionary pathways has been established.

The sequential deletions in the MTBC can be used to distinguish between strains and lineages. Many of the molecular typing protocols developed in clinical microbiology can be applied to ancient material, provided that the preservation is sufficiently good and the methods are adapted for short DNA fragments. The *M. tuberculosis* genome contains some repetitive sequences with specific regions that enable identification and diagnosis in clinical specimens. The first such region to be used was the insertion sequence IS*6110*, with a copy number normally between 1–25 per cell of *M. tuberculosis* (McEvoy *et al.*, 2007), but with only a single copy in *M. bovis*. The

alternative specific PCR, based on IS*1081* (Taylor *et al.*, 2005), is preferable in most cases, as there are normally six copies/cell in each member of the MTBC so quantification is possible. Optimisation of the PCR is necessary and modification of the PCR reaction mix is recommended for work with aDNA. Inclusion of stabilisers such as bovine serum albumin is often beneficial, probably due to a variety of effects such as masking non-specific binding sites, stabilising DNA fragments and binding or otherwise inactivating co-purified PCR inhibitors. Detection of amplicons has traditionally been based on agarose gel electrophoresis. Detection is also possible by hybridisation of labelled amplicons to a membrane, using a dot block technique, for example. However, real-time PCR enables the amplified product with an incorporated fluorescent marker or probe to be monitored directly via a computer screen and the methodology facilitates quantification.

M. tuberculosis strains can be distinguished from each other by spoligotyping – the analysis of unique spacer regions, each of 50–60 base pairs (bp), between identical Direct Repeat (DR) sequences. There is an international database and *M. tuberculosis* aDNA from many ancient specimens has been typed by this method. Spoligotyping can distinguish between members of the MTBC and has identified both *M. tuberculosis* and *Mycobacterium africanum* in Middle Kingdom (2050–1650 BC) ancient Egypt. In addition, this methodology enabled recognition of ancestral *M. tuberculosis* with an intact TbD1 region, and TbD1-deleted strains akin to those present in modern Europe and the Americas (Zink *et al.*, 2003, 2007). Early genotyping of *M. tuberculosis* aDNA was based on a combination of single nucleotide polymorphisms (SNPs), known as synonymous SNPs as there is no change in the amino acid coded, in the *katG* codon 463 (*katG*[463]), *gyrA* codon 95 (*gyrA*[95]) and DNA deletions. Later, Sreevatsan *et al.* (1997) re-defined these SNP types as principal genetic groups (PGGs).

1.3.   Early investigations of 18th-century tuberculosis in Hungary

Using conventional PCR and the molecular typing methods described above, historical tuberculosis was investigated in the 18th-century naturally mummified bodies from a sealed church crypt in the Hungarian town of Vác, north of Budapest (Fletcher *et al.*, 2003a, 2003b). During this period, tuberculosis was a common endemic disease in Central Europe, but had not reached the epidemic levels associated with the Industrial Revolution (Hutás, 1999). The preservation of these natural mummies was especially good, caused by the combination of low temperature, restricted oxygen levels, lack of any contact with soil or water, and the protection afforded by clothing and wood chips found in the closed pine coffins (Pap *et al.*, 1997, 1999). In addition to inscriptions on the coffins, there was both a church and civic archive that enabled identification of individuals, their family or community, and in some has been re-ordered and includes both cases, occupation and symptoms before death.

Tuberculosis was widespread in this population (Fletcher *et al.*, 2003a; Donoghue, 2011) and was analysed by age at death and family group. In one family of a mother and two daughters, the mother was infected with *M. tuberculosis* of PGG2, whereas her two daughters had *M. tuberculosis* of PGG3 (Fletcher *et al.*, 2003b). However, additional genetic markers suggested that each daughter was infected with a different *M. tuberculosis* strain.

1.4. New developments in genome analysis led by technological advances

PCR is a competitive reaction so it may not be possible to detect multiple genotypes in a DNA extract. However, recent developments in high-throughput sequencing and bioinformatics workflows have made it possible to sequence entire genomes directly, without DNA amplification. To overcome the problem of low aDNA yield, before performing next generation sequencing (NGS) an additional step can be inserted to enrich the target aDNA by using hybridisation capture, using modern sequences as bait on magnetic beads or an array. DNA fragments are then tagged by oligonucleotide ligation, sequenced and resulting sequence data assembled to construct the genome using bioinformatics software (Donoghue, 2013). This was the method used to determine the *M. tuberculosis* genome in a 19th-century skeleton from a church crypt in Leeds, Yorkshire, UK (Bouwman *et al.*, 2012). However, a potential problem with hybridisation capture is that the baits are based on modern references and so are not designed to detect ancestral sequences absent in modern strains. A more promising approach is 'metagenomics' – open-ended sequencing without targeted amplification or enrichment. The use of a metagenomics approach provides data on all DNA present in a sample, subject to DNA quality and fragment length. Therefore, the host aDNA should be present, plus all the microbes responsible for the natural decay process after death. This approach was used to study *Mycobacterium leprae* from the dental pulp region in a mediaeval case of leprosy (Schuenemann *et al.*, 2013), and to investigate the tuberculosis in one of the family members from Vác described above (Chan *et al.*, 2013). That particular study is the basis of the work described in detail below.

2. Materials and Methods

2.1. Source of samples examined

Human remains from the Vác church crypt are located in the Department of Anthropology, Hungarian Natural History Museum, Budapest, Hungary. Specimens were taken for analysis, using a no-touch technique with an endoscope that was de-contaminated between samples with detergent and 70% ethanol. Specimens were placed into sterile containers, each within a plastic bag and stored at 4 °C before transport to UCL. After preliminary screening using conventional PCR, it was found that tuberculosis was widespread in this population, including the three individuals described below. In the 2013 study,

molecular analysis was performed on samples from the following human remains:

Body 68: Terézia Hausmann, daughter of Johannis Hausmann, who died on December 25th 1797, aged 28 years. Material from the left chest was sampled (15–20 mg). There was no gross pathological or radiographic evidence of disease in the sample that was taken. Pulmonary tuberculosis was initially suspected for the following reasons; firstly, the mummified body was extremely small for the age of 28 years recorded in the archive, and was described as cachectic by clinicians, typical of someone suffering from tuberculosis. In addition, the archival record states that she died four years after her mother and less than three years after her younger sister, both of whom also had strong indications of tuberculosis.

Body 28: Annamária Schőner, wife of Johannis Hausmann, who died on December 16th, 1793, aged 55 years. Abdominal contents were sampled (17.3 mg). She was the mother of body 68. Her height was noted as 1.45 m but no other observations were made during sampling.

Body 121: László Beniczky, land-steward of the Bishopric of Vác, and Court Judge of Pest and Nógrád Counties, who died on November 7th 1764 aged 36 years. Rib bone scrapings (25 mg) were examined.

2.2. DNA extraction

Detailed methods have been published previously (Évinger *et al.*, 2011; Chan *et al.* 2013). Briefly, mineralised or mummified tissue was added to 400 µl of demineralisation solution (EDTA and Proteinase K) together with glass beads, and mixed on a bead beater. Samples were shaken at low speed for 48 hours at 56 °C. Covalent bonds can develop over time in aDNA, so to facilitate DNA strand separation by breaking cross-links, *N*-phenacythiazolium bromide (40 µl at 0.1M) was added and the mixture incubated at 56 °C for one hour. Lysis buffer (4.5 ml) based on guanidium thiocyanate was added, mixed and incubated for a further 72 hours at 56 °C. Each sample was subjected to three cycles of snap freezing in liquid nitrogen followed by thawing in a hot water bath to ensure complete disruption. Finally, samples were centrifuged for 15 min at 13 000 *g* and the supernatant collected in a sterile tube. Silica suspension (20 µl of 12% SiO$_2$ w/v, pH 2·0) was added to the supernatant and the mixture gently shaken for 1 hour at room temperature. The suspension was re-mixed, centrifuged for 1 min at 13 000 *g* and the supernatant discarded. The silica pellet was washed with buffer (200 µl of 5M GuSCN/Tris-HCL pH 6.4), twice with 200 µl of 70% (v/v) ethanol (-20 °C), once with 200 µl acetone (-20 °C), drained and left to dry in a 56 °C heating block. DNA was eluted with 80 µl elution buffer (Qiagen) at 60 °C for 15 min and stored at -20 °C.

2.3. Library preparation

The library preparation for aDNA from these Hungarian mummies is described in Chan *et al.* (2013) Supplementary Appendix and Kay *et al.* (2015).

2.3.1. Body 68

The fragment lengths of extracted DNA (100 μl) were analyzed on an Agilent Bioanalyzer and found to be at least 800 bp, with some as large as 4 kb. As a result samples were sheared using a Diagenode BioRuptor, for 30 cycles (30 seconds on/30 seconds off). This device uses pulsed ultrasound waves to create focused mechanical stress to fragment DNA. Using physical methods such as mechanical stress to fragment DNA produces heterogeneous ends –a mixture of 5' (positive strand) overhangs, 3' (negative strand of the double helix) overhands and blunt ends. In addition these overhangs will have varying lengths and may or may not be phosphorylated. Consequently the first step of library preparation was repair of the ends of DNA fragments using T4 DNA polymerase and KIenow enzyme. This step converts the overhangs into blunt ended fragments and T4 PNK was used to phosphorylate the 5' end.

To prevent the blunt-ended fragments from ligating to one another, the 3' end was adenylated using KIenow Exo-enzyme. This single base overhang is complementary to the single 'T' nucleotide on the 3' end of the adapter that is ligated to the fragment in the next step. In this step a mixture of two adapters (a 5' phosphorylated 20 bp adapter and a 33 bp adapter) were added to the 5' and 3' end of each strand. These adapters increase the fragment size of the final amplified library by 121 bp. The resulting products were purified with Agencourt AMPure XP beads (Beckman Coulter), which selectively binds DNA with a size-range between 300-600 bp allowing the removal of unligated adapters, adapters which have ligated to one another and products <300 bp and >600 bp. These four steps (end-repair, 3' adenylation, adapter ligation and purification) were performed on the Beckman SPRIworks Fragment Library System I (Beckman Coulter).

The size-selected fragments with adapter molecules on both ends were then enriched using 18 cycles of PCR (with Phusion DNA polymerase). At this stage the additional sequences are added to the ends of the adapter. At either end, the final amplified libraries now contain sequences that enable hybridisation to primers spotted on the flow cell surface, i.e. they allow the libraries to bind to the surface of the flow cell and sequencing primers from which sequencing-by-synthesis can begin. The final amplified libraries were purified to remove un-incorporated primers using Agencourt AMPure XP beads (Beckman Coulter) and analysed on an Agilent Bioanalyzer using a HS dsDNA chip. The final amplified library was found to be ~ 440 bp in length, as such the insert size was ~ 320 bp.

2.3.2. Bodies 28 and 121

Extracted DNA was not fragmented prior to library creation as fragment size was found to be <800 bp. Instead DNA eluted from silica pellets was used directly as input DNA for the NEBNext Ultra library prep kit. This protocol follows a similar process as that described for Body 68 with the following modifications – a single looped adapter was ligated to the end repaired, dA-tailed DNA fragment. After the ligation reaction a USER enzyme was added to cleave the uracil base in the loop of the looped adapter. Adaptor-ligated DNA fragments were not size-selected and in the PCR step, NEBNext high fidelity PCR master mix was used to enrich the final libraries over 15 cycles.

2.4. High-throughput sequencing

Final amplified libraries were diluted to 4nM and denatured with an equal volume of freshly made NaOH (0.1M). The single-stranded library was neutralized with LNA1 buffer from the Illumina NexteraXT kit, mixed and diluted to 20 pM. A volume of 600 μl was then loaded into an Illumina v2 2 × 150 bp kit for the library of body 68 and a v2 2 × 250 bp kit for the libraries from bodies 28 and 121.

The Illumina MiSeq carries out on-board cluster generation and sequencing-by-synthesis in 24 (for 2 × 150 bp kits) or 39 (for 2 × 251 bp kits) hours. In the machine, the adaptor sequence on the 5'ends of the single-stranded libraries bind to complementary oligonucleotides spotted the surface of a glass flow cell. The hybridised libraries are then amplified by means of PCR to create molecular clonal-colonies (MCC) on the flow cell. These are distributed in discrete clusters across the surface. After cluster generation the 3'end of the MCC is blocked and a universal sequencing primer (complementary to a portion of the adapter sequence) annealed to the MCCs. Fluorescent-labelled di-deoxynucleotides are sequentially incorporated; this is monitored by means of laser excitation of the fluorophore immediately after a base is incorporated. After each cycle (one cycle consists of addition of an A, T, C or G) the fluorophore is cleaved off and the terminator removed. Consequently the polymerase is now able to add another base in the next cycle of synthesis. The first read is complete after 151 or 251 cycles depending on the kit used. At the end of read one, the newly synthesised sequenced strand is stripped off and the 3'end of the MCC unblocked allowing it to anneal to complementary oligonucleotides (oligos) on the surface of the flow cell forming a bridge. The bridges are then linearised through the release of the MCC at the 5'end resulting in the inversion of the DNA strands. This is known as the turn-around. The second read proceeds in an identical fashion to the first read for a further 151/251 bp.

2.5. Metagenomic profiling of aDNA from body 68

Sequence reads from run one were mapped against the human genome (NCBI36) using Bowtie 2 to identify human DNA contamination (<1%) (Langmead and Salzberg, 2012). Remaining reads were assembled using CLC workbench (CLC bio, Denmark) to produce 72,887 contigs (total length = 44 Mb, N50 = 766). All assembled contigs were binned into genomes based on Z scores and coverage (Teeling *et al.*, 2004). Housekeeping genes (*dnaK*, *rpoB*, *gyrB* and *recA*) were identified in each genome using BLASTn. The contigs containing these genes were then compared against the nr dataset at NCBI using BLASTx to identify the closest match.

The quality of raw sequence files from run 2 was determined using FastQC (http://www.bioinformatics. babraham.ac.uk/projects/fastqc/). Reads were subsequently trimmed using fastq_quality_filter from the FAST-X toolkit with a minimum quality of 28 across 97% of the sequence. To ensure maintenance of paired reads the programme cmpfastq (http://compbio. brc.iop.kcl.ac.uk/software/cmpfastq.php), was used on the trimmed reads. Metagenomic reads were assembled using VelvetOptimiser with the following options: --ShortPaired --s 39 – e 119 – x 10 –t 4. An optimal Kmer of 95 was used for assembly of contigs meta-velvetg was run with an insert length of 196. Gene calling and annotation was achieved using prokka-1.6 (http:// www.vicbioinformatics.com/software.prokka.shtml) with the parameters: --metagenome ON and --mincontig 150. The resulting predicted proteins were used as an input for Phyla_AMPHORA (http://wolbachia.biology. virginia.edu/WuLab/Software.html), which was run with the following parameters: Phylum 0, WithReference, OutputFormat phylip and maximum likelihood. The resulting output was filtered to exclude results with less than 90% confidence and the remaining predicted proteins used to determine the phylotype of identified marker proteins in the metagenomic sample.

2.6. SNP calling on aDNA from body 68

Metagenomic reads from the first run were mapped against the H37Rv genome using BWA (Burrows-Wheeler Aligner – an algorithm for mapping low-divergent sequences against a large reference genome, such as the human genome) and single nucleotide polymorphisms (SNPs) called using Samtools (Li and Durbin, 2009; Li *et al.*, 2009). Creating random sub-sequences of 150 bases from the 7199/99 genome (HE663067) (Roetzer *et al.*, 2013) identified SNPs between 7199/99 and HR37Rv. Enough of these sub-sequences were generated to simulate 40-fold coverage. The 'pseudo reads' were then mapped against HR37Rv and SNPs identified by the methods described. SNPs were filtered with a minimum and maximum coverage of 5 and 80 respectively and a minimum mapping score (assigned by Samtools) of 100. SNPs were compared, filtered, their gene location and resulting amino acid change calculated and the underlying reads viewed manually using a custom Java program.

Reads from run 2 were mapped against the reference genome *Mycobacterium tuberculosis* 7199/99 (NC_020089.1) using Bowtie2 (Langmead and Salzberg, 2012) with very-sensitive parameter. SNPs were identified with Samtools mpileup command. The resulting output was further analysed with VarScan (Koboldt *et al.*, 2009), with the following parameters: --min-coverage 50, --min-reads 240, --min-avg-qual 28 and --p-value 0.05. Bases passing these filters were called as SNPs. Previous analysis (Chan *et al.*, 2013), had shown the presence of

multiple strains of *Mycobacterium tuberculosis* to be present in the sample. The resulting SNPs from these data again suggested the presence of multiple strains of *M. tuberculosis*. To further analyse this, all SNPs that occurred on the same sequencing read pair as another SNP were identified, thus being from the same original sample of DNA. The frequency of all pairs of SNPs was then calculated to infer the proportion of the two different *M. tuberculosis* genomes within the sample.

3. Results

3.1. Evidence of M. tuberculosis *in Vác mummy samples*

In the first metagenomic analysis of the remains of Terézia Hausmann, body 68, 5.5 million paired-end reads were obtained using the Illumina v1 2 × 150 bp kit, and deposited in the Sequence Read Archive with the accession number SRP018736 (Chan *et al.*, 2013). Of these, 8% aligned against the *M. tuberculosis* reference strain H37Rv, giving an average depth of coverage of 32 × the total genome, with a very even spread of reads (Figure 5.1A). In contrast, less than 1% of the reads aligned against the human genome. Figure 5.1B Illustrates the very few if any reads mapped to the H37Rv locus 3378001–3380439. This suggests that these genes are absent in body 68. Analysis of SNPs and indels indicated that there was a mixed population of two *M. tuberculosis* genotypes in the sample, present in roughly equal proportions (Figures 5.1C and 5.1D). Both genotypes belong to the Haarlem lineage and are closest, among genome-sequenced strains, to the modern German strain of *M. tuberculosis*, 7199/99, which represents the reference genome for a tuberculosis outbreak that occurred in northern Germany from 1998 until 2010. Mummy strain M2 is more similar to 7199/99 than is mummy strain M1.

A second DNA extract from body 68 was prepared using a different piece of lung tissue. This time 19 million reads were obtained using the Illumina v2 2 × 251 bp kit. However, the proportion of reads aligning to the human genome was still <1% of total reads, with 28% of reads aligning to the *M. tuberculosis* 7199/99 genome sequence, giving 550-fold coverage. Metagenomics sequencing of samples from bodies 121, 28 and five other individuals from the crypt was also performed, but analysis of the sequences was completed only in 2015 (Kay *et al.*, 2015).

3. 2 Analysis of the majority microflora aDNA *in mummy 68*

The most abundant species (*c*. 200 times coverage) showed similarity to sequenced *Nocardia* species whilst the least abundant (*c*. ten times coverage) showed limited similarity to *Thermobifida fusca*. However, collectively these three genomes accounted for only around 25 % of the total assembled contigs, as the remainder were short <1000 bp with low coverage (>5). In run one, searching for *rpoB* genes amongst the assembled contigs revealed sequences similar to soil-dwelling spore forming genera

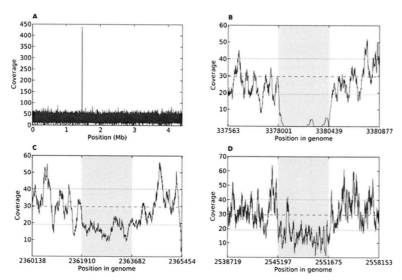

Figure. 5.1 A. Metagenomic reads from body 68 mapped against H37Rv, with deep and even coverage – the spike around position 1.5 Mb corresponds to the 16S rRNA operon; B. Illustration of the very few/no reads mapped to the H37Rv locus 3378001-3380439 suggesting these genes are not present in body 68 M. tuberculosis *genome(s); C. and D. The region corresponding to H37Rv locus 2361910-2363682 (1C) and H37Rv locus 2545197-2551675 (1D) – coverage was around half the average suggesting only one of two body 68 strains contains these deletions.*

Table 5.1 Best BLASTx match to assembled contigs containing the rpoB, recA, gyrB and dnaK genes.

Size of assembled contig	Coverage of contig	Closest BLASTx match	% id
84,468	32	*Mycobacterium tuberculosis H37Rv*	99
12743	231	*Nocardia brasiliensis ATCC 700358*	93
13529	13.8	*Thermobifida fusca YX*	84
1222	4.7	*Brachybacterium faecium DSM 4810*	92
1187	4.5	*Clostridium tetani E88*	88
692	5.1	*Clostridium acidurici 9a*	87
419	3.4	*Clostridium celatum DSM 1785*	90
392	6	*Brevibacterium iodinum*	88
321	1.8	*Bacillus anthracis*	97
514	6.2	*Finegoldia magna ATCC 29328*	82
714	6.4	*Brachybacterium faecium DSM 4810*	95
2299	9	*Clostridium acidurici 9a*	71
278	2.2	*Bordetella holmesii*	91

Bacillus and *Clostridium* (Table 5.1). The overall analysis of the non-mycobacterial microorganisms from the second metagenomic sample in mummy 68 showed that the Actinobacteria overwhelmingly predominate, followed by the Firmicutes. Both these phyla are common in the environment and play an important role in natural decay in the carbon cycle.

4. Discussion

4.1. Evidence of differential preservation of mycobacterial and human aDNA

This study illustrates the excellent aDNA preservation in body 68, that of Terézia Hausmann, and this was confirmed in a subsequent investigation of body 68 material where the average read length was found to be 277 bp (Kay *et al.*, 2015). The environmental conditions in the sealed church crypt in Vác were clearly ideal for the preservation of *M. tuberculosis* aDNA. Due to the thick, hydrophobic bacterial cell wall, *M. tuberculosis* aDNA is likely to be better preserved than the host human aDNA. This has been proposed in the past (Spigelman and Donoghue, 2003) and it has been demonstrated recently that mycobacterial DNA survives heat treatment better than human DNA (Nguyen-Hieu *et al.*, 2012).

A recent study of historical *Mycobacterium leprae* samples also demonstrated better preservation of mycobacterial DNA than of human DNA. In all but one of the samples from that study, *M. leprae* sequences showed damage motifs below 10%, while the associated human mitochondrial aDNA had >20% damaged templates (Scheunemann *et al.*, 2013). The authors supported the view that the better preservation of mycobacterial DNA was possibly due to a waxy cell wall, as suggested previously (Spigelman and Donoghue, 2003; Donoghue *et al.*, 2004), but they were more concerned that this prevented damage motifs from being used as a criterion of authenticity of *M. leprae* aDNA.

4.2. Implications of a mixed infection with two strains of M. tuberculosis

The fact that Terézia Hausmann was infected with two strains of *M. tuberculosis*, suggests that the disease was hyperendemic in 18th-century Vác – a phenomenon that can be found today in parts of the globe with high levels of tuberculosis. Prior infection does not protect against acquisition of another strain (Cohen *et al.*, 2012) and re-infection plays a dominant role in disease dynamics when the disease prevalence increases past a threshold. Therefore, the presence of a mixed infection in 18th-century Hungary illustrates the disease burden that was experienced by this population.

The modern *M. tuberculosis* strain 7199/99, that resembles the 18th-century Hungarian strains, is in the virulent Erdman lineage. This clone was responsible for a large outbreak that lasted over 11 years in over 2000 patients in Hamburg, Germany. In comparison with other strains of *M. tuberculosis* in this region, it had a faster growth rate and was transmitted more readily. In a follow-up study, we obtained further evidence of the genomic epidemiology of tuberculosis in 18th-century Hungary (Kay *et al.*, 2015). This demonstrated that at least 14 sub-lineages of *M. tuberculosis* were present in this population and that mixed infections were widespread. A similar scenario can be seen today in sub-Saharan Africa, highlighting the importance of high population density of individuals in tuberculosis infections (Cohen *et al.*, 2012).

Acknowledgements

The Leverhulme Trust Project Grant F/00 094/BL (OY-CL, DEM, GSB). GSB was supported by a Personal Research Chair from Mr. James Bardrick and the UK Medical Research Council.

References

Allison, M.J., Mendoza, D., Pezzia, A. 1975. Documentation of a case of tuberculosis in Pre-Columbian America. *American Review of Respiratory Disease* 107(6), 985–991.

Barnes, I., Duda, A., Pybus, O.G., Thomas, M.G. 2010. Ancient urbanization predicts genetic resistance to tuberculosis. *Evolution* 65(3), 842–848.

Blaser, M.J., Kirschner, D. 2007. The equilibria that allow bacterial persistence in human hosts. *Nature* 449(7164), 843–849.

Bouwman, A.S., Kennedy, S.L., Müller, R., Stephens, R.H., Holst, M., Caffell, A.C., Roberts, C.A., Brown, T.A. 2012. Genotype of a historic strain of *Mycobacterium tuberculosis*. *Proceedings of the National Academy of Sciences of the United States of America* 109(45), 18511–18516.

Brosch, R., Gordon, S.V., Marmiesse, M., Brodin, P., Buchrieser, C., Eiglmeier, K., Garnier, T., Gutierrez, C., Hewinson, G., Kremer, K., Parsons, L., Pym, A.S., Samper, S., van Soolingen, D., Cole, S.T. 2002. A new evolutionary scenario for the *Mycobacterium tuberculosis* complex., *Proceedings of the National Academy of Sciences of the United States of America* 99(6), 3684–3689.

Chan, J.Z-M., Sergeant, M.J., Lee, O. Y-C., Minnikin, D.E., Besra, G., Pap, I., Spigelman, M., Donoghue, H.D., Pallen, M.J. 2013. Metagenomic analysis of tuberculosis in a mummy. *The New England Journal of Medicine* 369(3), 289–290 and Suppl. Appendix. doi: 10.1056/NEJMc1302295

Cohen, T., van Helden, P.D., Wilson, D., Colijn, C., McLaughlin, M.M., Abubakar, I., Warren, R.M. 2012. Mixed-strain *Mycobacterium tuberculosis* infections and the implications for tuberculosis treatment and control. *Clinical Microbiology Reviews* 25(4), 708–719.

Comas, I., Coscolla, M., Luo, T., Borrell, S., Holt, K.E., Kato-Maeda, M., Parkhill, J., Malla, B., Berg, S., Thwaites, G., Yeboah-Manu, D., Bothamley, G., Mei, J., Wei, L., Bentley, S., Harris, S.R., Niemann, S., Diel, R., Aseffa, A., Gao, Q., Young, D., Gagneux, S. 2013. Out-of-Africa migration and Neolithic coexpansion of *Mycobacterium tuberculosis* with modern humans. *Nature Genetics* 45(10), 1176–1182. doi: 10.1038/ng.2744.

Corthals, A., Koller, A., Martin, D.W., Rieger, R., Chen, E.I., Bernaski, M., Recagno, G., Dávalos, L.M. 2012. Detecting the immune system response of a 500 year-old Inca mummy. *PLoS ONE* 7(7), e41244.

Donoghue, H.D. 2008. Palaeomicrobiology of tuberculosis, in: Raoult, D., Drancourt, M. (Eds.), *Paleomicrobiology. Past Human Infections*. Springer-Verlag, Berlin and Heidelberg, pp. 75–97.

Donoghue, H.D., Pap, I., Szikossy, I., Spigelman, M. 2011. Detection and characterization of *Mycobacterium tuberculosis* DNA in 18th century Hungarians with pulmonary and extra-pulmonary tuberculosis, in: Gill-Frerking, G., Rosendahl, W., Zink, A., Piombino-Mascali, D. (Eds.), *Yearbook of Mummy Studies 1*. Verlag Dr. Friedrich Pfeil, München, Germany, pp. 51–56.

Donoghue, H.D. 2013. Insights into ancient leprosy and tuberculosis using metagenomics. *Trends in Microbiology* 21(9), 448–450.

Donoghue, H.D., Spigelman, M., Zias, J., Gernaey-Child, A.M., Minnikin, D.E. 1998. *Mycobacterium tuberculosis* complex DNA in calcified pleura from remains 1400 years old. *Letters in Applied Microbiology* 27(5), 265–269.

Donoghue, H.D., Spigelman, M., Greenblatt, C.L., Lev-Maor, G., Kahila Bar-Gal, G., Matheson, C., Vernon, K., Nerlich, A.G., Zink, A.R. 2004. Tuberculosis: From prehistory to Robert Koch, as revealed by ancient DNA. *Lancet Infectious Diseases* 4(9), 584–592.

Évinger, S., Bernert, ZS., Fóthi, E., Wolff, K., Kővári, I., Marcsik, A., Donoghue, H.D., O'Grady, J., Kiss, K.K., Hajdu, T. 2011. New skeletal tuberculosis cases in past populations from Western Hungary (Transdanubia). *HOMO – Journal of Comparative Human Biology* 62(3), 165–183.

Fletcher, H.A., Donoghue, H.D., Holton, J., Pap, I., Spigelman, M. 2003a. Widespread occurrence of *Mycobacterium tuberculosis* DNA from 18th–19th century Hungarians. *American Journal of Physical Anthropology* 120(2), 144–152.

Fletcher, H.A., Donoghue, H.D., Taylor, G.M., van der Zanden, A.G.M., Spigelman, M. 2003b. Molecular analysis of *Mycobacterium tuberculosis* DNA from a family of 18th century Hungarians. *Microbiology* 149(1), 143–151.

Gagneux, S., Deriemer K., Tran, V., Kato-Maeda, M., De Jong, B.C., Narayanan, S., Nicol, M., Niemann, S., Kremer, K., Gutierrez, M.C., Hilty, M., Hopewell, P.C., Small, P.M. 2006. Variable host-pathogen compatibility, in: *Mycobacterium tuberculosis*. *Proceedings of the National Academy of Sciences of the United States of America* 103(8), 2869–2873.

Gutierrez, M.C., Brisse, S., Brosch, R., Fabre, M., Omaïs, B., Marmiesse, M., Supply, P., Vincent, V. 2005. Ancient origin and gene mosaicism of the progenitor of *Mycobacterium tuberculosis*. *PLoS Pathogens* 1(1), e5.

Holloway, K.L., Link, K., Rühli, F., Henneberg, M. 2013. Skeletal lesions in human tuberculosis may sometimes heal: An aid to palaeopathological diagnoses. *PLoS ONE* 8(4), e62798. doi: 10.1371/journal.pone.0062798

Kay, G.L., Sergeant, M.J., Zhou, Z., Chan, J.Z.-M., Millard, A., Quick, J., Szikossy, I., Pap, I., Spigelman, M., Loman, N.J., Achtman, M., Donoghue, H.D., Pallen, M. 2015. Eighteenth-century genomes show that mixed infections were common at time of peak tuberculosis in Europe. *Nature Communications* 6(6717). doi: 10.1038/ncomms7717

Koboldt, D.C., Chen, K., Wylie, T., Larson, D.E., McLellan, M.D., Mardis, E.R., Weinstock, G.M., Wilson R.K., Ding, L. 2009. VarScan: Variant detection in massively parallel sequencing of individual and pooled samples. *Bioinformatics* 25(17), 2283–2285.

Langmead, B., Salzberg, S.L. 2012. Fast gapped-read alignment with Bowtie 2. *Nature Methods* 9(4), 357–359.

Lee, O.Y.-C., Wu, H.H.T., Donoghue, H.D., Spigelman, M., Greenblatt, C.L., Bull, I.D., Rothschild, B.M., Martin, L.D., Minnikin, D.E., Besra, G.S. 2012. *Mycobacterium tuberculosis* complex lipid virulence factors preserved in the 17,000-year-old skeleton of an extinct bison, *Bison antiquus*. *PLoS ONE* 7(7), e41923.

Li, H., Durbin, R. 2009. Fast and accurate short read alignment with Burrows–Wheeler transform. *Bioinformatics* 25(14), 1754–1760.

Li, H., Handsaker, B., Wysoker, A., Fennell, T., Ruan, J., Homer, N., Marth, G., Abecasis, G., Durbin, R., 1000 Genome Project Data Processing Subgroup. 2009. The sequence alignment/map format and SAMtools. *Bioinformatics* 25(16), 2078–2079.

McEvoy, C.R., Falmer, A.A., Gey van Pittius, N.C., Victor, T.C., van Helden, P.D., Warren, R.M. 2007. The role of IS*6110* in the evolution of *Mycobacterium tuberculosis*. *Tuberculosis* (Edinburgh) 87(5), 393–404.

Minnikin, D.E., Lee, O.Y.-C., Wu, H.H.T., Besra, G.S., Donoghue, H.D. 2012. Molecular biomarkers for ancient tuberculosis, in: Cardona, P.-J. (Ed.), *Understanding tuberculosis – Deciphering The Secret Life of The Bacilli.* InTech Open Access Publisher, Rijeka, Croatia, pp. 3–36. Available from: http://www.intechopen.com/articles/ show/title/molecular-biomarkers-for-ancient-tuberculosis (Accessed 17th February, 2012)

Nguyen-Hieu, T., Aboudharam, G., Drancourt, M. 2012. Heat degradation of eukaryotic and bacterial DNA: An experimental model for paleomicrobiology. *BMC Research Notes* 5, 528. doi: 10.1186/1756-0500-5-528.

Pap, I., Józsa, L., Repa, I., Bajzik, G., Lakhani, S.R., Donoghue, H.D., Spigelman, M. 1999. 18–19th century tuberculosis in naturally mummified individuals (Vác, Hungary), in: Pálfi, G., Dutour, O., Deák, J., Hutás, I. (Eds.), *Tuberculosis Past and Present.* Golden Book Publisher Ltd., Tuberculosis Foundation, Budapest, pp. 421–428.

Pap, I., Susa, É., Józsa, L. 1997. Mummies from the 18th–19th century Dominican church of Vác, Hungary. *Acta Biologica Szegediensis* 42, 107–112.

Roetzer, A., Diel, R., Kohl, T.A., Rückert, C., Nübel, U., Blom, J., Wirth, T., Jaenicke, S., Schuback, S., Rüsch-Gerdes, S., Supply, P., Kalinowski J., Niemann, S. 2013. Whole genome sequencing versus traditional genotyping for investigation of a *Mycobacterium tuberculosis* outbreak: A longitudinal molecular epidemiological study. *PLoS Medicine* 10(2), e1001387.

Salo, W.L., Aufderheide, A.C., Buikstra, J., Holcomb, T.A. 1994. Identification of *Mycobacterium tuberculosis* DNA in a pre-Columbian Peruvian mummy. *Proceedings of the National Academy of Sciences of the United States of America* 91(6), 2091–2094.

Schuenemann, V.J., Singh, P., Mendum, T.A., Krause-Kyora, B., Jäger, G., Bos, K.I., Herbig, A., Economou, C., Benjak, A., Busso, P., Nenbel, A., Boldsen, J.L., Kjellström, A., Wu, H., Stewart, G.R., Taylor, G.M., Bauer, P., Lee, O.Y.-C., Wu, H.H.T., Minnikin, D.E., Besra, G.S., Tucker, K., Roffey, S., Sow, S.O., Cole, S.T., Nieselt, K., Krause, J. 2013. Genome-wide comparison of Medieval and modern *Mycobacterium leprae*. *Science* 341(6142), 179–183 and Supplement.

Smith, N.H., Hewinson, R.G., Kremer, K., Brosch, R., Gordon, S.V. 2009. Myths and misconceptions: The origin and evolution of *Mycobacterium tuberculosis*. *Nature Reviews Microbiology* 7(7), 537–544.

Spigelman, M. and Donoghue, H.D. 2003. Palaeobacteriology with special reference to pathogenic mycobacteria, in: Greenblatt, C., Spigelman, M. (Eds.), *Emerging Pathogens: Archaeology, Ecology and Evolution of Infectious Disease.* Oxford University Press, Oxford, pp. 175–188.

Spigelman, M., Lemma, E. 1993. The use of the polymerase chain reaction (PCR) to detect *Mycobacterium tuberculosis* in ancient skeletons. *International Journal of Osteoarchaeology* 3(2), 137–143.

Sreevatsan, S., Pan, X., Stockbauer, K. E., Connell, N. D., Kreiswirth, B. N., Whittam, T. S., Musser, J. M. 1997. Restricted structural gene polymorphism in the *Mycobacterium tuberculosis* complex indicates recent global dissemination. *Proceedings of the National Academy of Sciences of the United States of America* 94(18), 9869–9874.

Taylor, G.M., Young, D.B., Mays, S.A. 2005. Genotypic analysis of the earliest known prehistoric case of tuberculosis in Britain. *Journal of Clinical Microbiology* 43(5), 2236–2240.

Teeling, H., Meyerdierks, A., Bauer M., Amann R., Glöckner, F.O. 2004. Application of tetranucleotide frequencies for the assignment of genomic fragments. *Environmental Microbiology* 6(9), 938–947.

Tran T.-N.-N., Aboudharam G., Raoult D., Drancourt M. 2011. Beyond ancient microbial DNA: nonnucleotidic biomolecules for palaeomicrobiology. *BioTechniques* 50, 370–380. doi: 10.2144/000113689

Zink, A.R., Grabner, W., Reischl, U., Wolf, H., Nerlich, A.G. 2003. Molecular study on human tuberculosis in three geographically distinct and time delineated populations from ancient Egypt. *Epidemiology and Infection* 130(2), 239–249.

Zink, A.R., Molnár, E., Motamedi, N., Pálfi, G., Marcsik, A., Nerlich, A.G. 2007. Molecular history of tuberculosis from ancient mummies and skeletons. *International Journal of Osteoarchaeology* 17(4), 380–391.

6

From Soldiers to Military Communities in Roman London: A Bioarchaeological Perspective

Rebecca Redfern

The army is a key research area in Roman archaeology, reflecting both the historical development of the discipline but also the wealth of primary and secondary evidence for this force of Imperial authority. The majority of work concentrates on material culture and inscription evidence; bioarchaeology has yet to make significant in-roads to its study. In response, this paper seeks to address two aspects of the debate: can bioarchaeology be used to identify Roman military personnel, and what contribution can bioarchaeology make to the growing body of data for the presence of families and slaves in settlements with a strong military connection.

Keywords Household; Slavery; Bioarchaeology; Army; Roman

1. Introduction

Roman soldiers had one of the most 'visible' identities in the Roman Empire (James, 1999, 2011a). In life, serving members of the army and auxiliary troops had a distinctive material culture (e.g., armour, weaponry and dress accessories) and dietary practices (Webster, 1998; Goldsworthy and Haynes, 1999; James, 1999; King, 1999; Cool, 2006; Erdkamp, 2007). In death, military personnel were one of the main groups who practiced the 'epigraphic habit', whereby a person's name, place of origin and occupation were recorded on their funerary monument (Saller and Shaw, 1984; Hope, 1997a; Bodel, 2001) – there is also inscription evidence to show that groups of soldiers were buried together, maintaining the strong bonds and social groupings (*collegia*) created during service (MacMullen, 1984; Bishop, 1990; Carroll, 2006; Haynes, 2013). The primary sources concerning soldiers range from personal and professional documents written by serving legionaries (i.e. Bowman, 2004) to a variety of works about them, including satirical poetry and theatrical works (i.e. Juvenal, 2008), accounts of military campaigns and activities (i.e. Caesar, 1983), as well as descriptions of their daily activities (i.e. Josephus, 1981).

The army has been a popular research area from the earliest days of Roman archaeology, not only because of the wealth and diversity of archaeological and primary source evidence, but also because many of the individuals working in the field came from a military background (Hingley, 2000, 2005, 2009). This gave rise to a view of the army that emphasised violence, masculinity, weaponry and tactics (James, 2002). However, since the integration of feminist and gender theories in archaeology, the accumulated evidence for the military has been revised, and now interpretations show that forts were not solely male or military living spaces (Roxan, 1991; Allison *et al.*, 2005; James, 2006). These changes in approach, have allowed us to recognise that the army was a diverse organisation, greater than the legionaries and auxiliary troops in service, one which was supported and facilitated by a wider community, made up of many disparate groups of people (James, 1999; Erdkamp, 2007; Haynes, 2013). One important group were military slaves, who were state-owned and subject to military discipline, they played an important logistical role both on campaign and at base, because the army did not have contracts with merchants to supply them (Roth and Roth, 1998); soldiers were able to own slaves and were often responsible for capturing them on campaign (Scheidel, 1997). There were also freedmen and other labourers or servants who were attached to the army to undertake various manual jobs (Roth and Roth,

1998). Examination of the material culture excavated from forts has shown that these were also inhabited by women and children (and their family slaves), many of whom were families of the garrison commanders, explicitly evidenced by the letters between such wives discovered at Vindolanda (England) (Bowman, 2004). Many fortresses had settlements known as *canabae* located nearby. These civilian communities were populated by the wives and families of serving soldiers, itinerant workers, and traders who sold additional supplies, equipment, entertainment, and provided anything else the army may have required (Roth and Roth, 1998; Goldsworthy and Haynes, 1999; James, 2002; Erdkamp, 2007; Phang, 2008). Initially, soldiers were not allowed to be married but by the 4th century AD, combined with longer-spells of garrison life, the laws changed and permitted them to have legitimate families (Scheidel, 2005; Knapp, 2011). Many tombstones from across the Empire were commissioned by the wives (legal or not) of military personnel, many of whom had begun as their slaves (Knapp, 2011). There is also evidence for serving military personnel having same-sex relationships with their male slaves, other soldiers, male prostitutes and men in the local communities (Phang, 2008).

After at least 20 years of service, discharged soldiers received their diploma (Roxan and Holder, 2003), documents which also list their family members, such as wives, sisters and children. Many veterans were also given a grant of land in the provinces where they settled, in a colony (*colonia*) where groups of veterans (e.g. *collegia*) would live in entire blocks, or else in a *canabae* (often with their family), where they established themselves as farmers or as business men, supplying goods and services to local military units (MacMullen, 1984; Garnsey and Saller, 1987; Keppie, 2000). By the conquest of Britain, the majority of the army were recruited from outside Italy and it seems that many chose to retire where they served, although many from the auxiliary units returned to their homelands (Mann, 2002; Haynes, 2013).

Human remains have been recovered from military locales, ranging from disarticulated pieces of bone, infant burials, ritual burials, formal cremation and inhumation burials of children and adults, to the mass internment of fallen legions (e.g., in Britain, Wells, 1957–8; Bennett *et al.*, 1982; Cool, 2004; Wells, 2004; Croom and Caffell, 2005; Wilmott *et al.*, 2009; Booth, 2014). In the Empire, the remains of a handful of individuals have been encountered in association with military tombstones, and in some very rare instances, men still dressed in their military equipment have been discovered, many of whom suffered a violent death (see section 2) (Constandse-Westermann, 1982; Andersen, 2009; Golubović *et al.*, 2009; James, 2011b). Cemeteries associated with military settlements have also been excavated, such as those along Hadrian's Wall (England) and at Vindonissa (Switzerland), which contained the burials of subadults, males and females (Cool, 2004; Carroll, 2012). Interestingly, these do not always have a male-biased demography, such as the

cremation cemetery from Brougham (Cumbria, England) which had more females present (McKinley, 2004).

Bioarchaeology is a growing area in Roman studies, despite many years of scepticism by archaeologists, classicists and ancient historians (i.e. Scheidel, 1996, 2010). Its approaches and methods (Buikstra and Beck, 2006; Knudson and Stojanowski, 2008) allow it to make a valuable contribution to understanding the impact of urbanisation, colonisation, gender, life course and status, with stable isotopes making a significant impact on our knowledge of diet and mobility (e.g., Dupras *et al.*, 2001; Gowland, 2001, 2006; Gowland and Chamberlain, 2002; Prowse *et al.*, 2004, 2007; Cucina *et al.*, 2006; Craig *et al.*, 2009; Eckardt, 2010, 2014; Redfern and DeWitte, 2011; Redfern and Gowland, 2011; Redfern *et al.*, 2012; Killgrove, 2013; Powell *et al.*, 2014).

In order to investigate whether members of the military and its community could be identified in the cemeteries of Roman London (*Londinium*), this paper addresses two aspects of the wider debate outlined above: can bioarchaeology be used to identify Roman military personnel? And what contribution can bioarchaeology make to the growing body of data for the presence of families and slaves in settlements, particularly those associated with the military.

2. Addressing the Problem
2.1. *Funerary perspectives*

The military were one of the largest exponents of epigraphy, particularly in death with monuments erected by their *collegia* (Hope, 1997a, 2003; Carroll, 2006). Soldiers who died before receiving their discharge were not buried with their military equipment; instead their heirs received its monetary value (Bishop, 2011). Consequentially, in the absence of tombstones, identifying military personnel is highly problematic (see Mrav, 2013). This issue was investigated by Jones (1984) who concluded that it was too difficult a task, and systematically critiqued all of the potential criteria that could be used (i.e. weapons and dress accessories). More recently, Andersen (2009) has tackled the question and made the following observations: there is no one form of archaeological evidence that indicates a military status; cemeteries do not appear to have been exclusively military or civilian but rather a mix of the two; cemeteries or burials near Roman forts contain very few military items; cremation burials may reflect the conservative nature of the military; military burials are more likely to be low status; and the burial of veterans will be harder to identify than serving soldiers. Her conclusions offer slightly more opportunities for research, because they raise the possibility that we should expect cemeteries associated with settlements inhabited by the military population to contain military men (in service or retired), and to perhaps target low-status and cremation burials.

Roman Britain was inhabited by many diverse communities, whose cultural identities were often incorporated into their funerals – a fact noted as far back as the 19th century (Davis and Thurnam, 1865). Epigraphic evidence from Britain very clearly shows that people came from the Mediterranean, North Africa and Continental Europe (Hope, 1997b; Rowland, 1976). Work by, amongst others, Cool (2010), Pearce (2010) and Swift (2009) has addressed the connections between cultural identity and origin in Britain and emphasised that any burial with imported material culture must be set in its wider context and in comparison to other cemetery populations of the same period. Their work, in combination with stable isotope studies, has shown that in many cases, people who had migrated to *Britannia* from other parts of the Empire were buried with dress accessories and other material culture reflecting and echoing their place of origin (Evans *et al.*, 2006; Eckardt *et al.*, 2009, 2014). For example, Pearce's (2010) work on the Eastern Cemetery of Roman London, suggests that a number of late Roman female burials with personal ornaments and value items could be the female counterpart to the male military 'aristocracy' burials seen in one male from the cemetery (see section 4.1), and also in the frontier (*limes*) cemeteries of Continental Europe (see also Cool, 2010). Pearce (2010) stresses the dynamic nature of funerary practices, in that the act of migration and the burying community are not static variables, and that it may be easier to identify (free) people who travelled long distances rather than local or regional migrations; these issues are exemplified by two publications. Prowse and colleagues (2010) studied the population buried in low-status, often unfurnished graves, from a large Imperial villa complex in southern Italy. They employed aDNA and stable isotope analyses and found connections to sub-Saharan Africa, East Asia and other parts of the Mediterranean, but none of the individuals who had migrated to Italy were buried with material culture reflecting this journey or non-'Roman' cultural affiliations. The study by Leach *et al.* (2010) which examined a 4th century AD high-status female burial from York, displayed a mixture of non-local and local material culture, and is believed to reflect her wider family and other (historical) connections. Estimation of this female's ancestry suggested that she was of mixed descent as her skull displayed morphological characteristics observed in both black and white populations. The stable isotope results suggested that some years before she died, she had migrated to York from western Britain or Europe, or the Mediterranean. She was buried in a stone coffin with jet and elephant ivory bangles, a glass mirror, a blue glass jug, jewellery and an openwork mount, probably for a box, which says, 'Hail, sister, may you live in God' (Leach *et al.*, 2010).

2.2. Stable Isotope Evidence

The growth of stable isotope studies in Roman bioarchaeology has provided a fresh perspective and enabled several studies to challenge and expand, often embedded, interpretations of ethnicity and cultural artefacts, such as those associated with 'Germanic' people in late Roman Britain (e.g. Evans *et al.*, 2006; Eckardt *et al.*, 2009, 2014). The extensive use of cremation in

Britain and Continental Europe, prevents its application to many burials (Schmidt and Symes, 2008), and therefore our knowledge and interpretation are skewed towards inhumation cemeteries, particularly those from southern Britain (Pearce, 2008). The majority of studies focus on individual cemeteries, with very few regional or settlement studies (Müldner, 2005, 2013; Cummings, 2008; Redfern *et al.*, 2010; Cheung *et al.*, 2012).

The evidence from military settlements support the epigraphic and material culture evidence for legions from North Africa and Continental Europe being stationed in *Britannia*, as individuals with isotopic signatures from these locales have been identified at different sites in York and northern England (Eckardt *et al.*, 2010). Chenery *et al.* (2011) sampled burials from the fort and associated civilian settlements at Catterick (Yorkshire). This study found that many people were locally born but a child and one female had originated from the Mediterranean, and these results suggest that by the late 4th century AD, army recruitment was local rather than Continental (Chenery *et al.*, 2011).

The value of stable isotope analysis to Roman bioarchaeology cannot be underestimated, particularly for burials without grave-goods (see section 2.1). It provides a unique opportunity to explore status groups who are generally 'invisible' in the archaeological record of the northwest provinces, particularly the enslaved (see Webster, 2005, 2010). However, it is recognised that this technique is not a 'magic bullet' (Pollard, 2011), it is limited to individuals with the requisite dentition and bone(s) present and impacted by funding constraints and permissions from the curating institution.

2.3. Bioarchaeology: An Osteobiographical Approach?

Several primary sources describe the physical characteristics of soldiers. The most comprehensive is a late Roman military handbook written by Renatus Vegetius (1993) (late 4th to early 5th century AD). It includes information about discipline, combat training and a list of the attributes that a recruit should have: be from a rural environment so that they will have been exposed to all types of weather and brought up doing hard labour; an adolescent, because instructions are more likely to be learnt - many enlisted at the age of 13 years old; to be quite tall (5 Roman feet and 7 inches) but that was not as important as strength; to have had a physically demanding job, like a carpenter; and to be trustworthy as they embodied the ethics and prowess of Imperial power (Davies, 1989; Phang, 2008). James (1999) reminds us that joining the military as a soldier would have created a fundamental shift in a man's identity/identities and body (see also Fowler, 2004; Insoll, 2007). The recruit's appearance and expressions of masculinity would have changed in order to conform with the community, though military exercise, hygiene and grooming practices (James, 1999); factors which would have impacted on their body because of its dynamic and plastic nature (Sofaer, 2006; Knudson and Stojanowski, 2009; Gowland and Thompson, 2013).

Table 6.1 Bioarchaeological summaries of published 'known' Roman military personnel (male).

Site and date	Burial context	Military equipment	Age in years	Stature (cm)	Pathology	Reference
Velsen (Holland) AD15–30	Wooden well: laid on his back, with his knees flexed and his head rolled forward onto his chest. The body was covered by 20 kg of stone and 60kg of broken quernstones, pottery and domestic refuse	Iron dagger without a handle; decorated scabbard; belt plates; bronze fibula; decorative button; bronze ring and the leather sole of a shoe.	25	190	A peri-mortem blunt force injury to the right frontal bone. Osteoarthritic changes to the spine, feet and right elbow joints and poor dental health (carious lesions and calculus)	Constandse-Westermann, 1982; Morel and Bosman, 1989
Canterbury Castle (Britain) AD mid-2nd to early 3rd centuries	Grave pit cut through the western edge of early Roman ditches	Sword and scabbard	≥30	–	Osteoarthritis of the lumbar vertebrae; exostosis at the 'lower end' of the right femur; healed fracture to the right fibula. Post-mortem fractures to ribs, left humerus, ulna and femur	Bennett et al., 1982; Garrard, 1982
		Sword, scabbard and hobnailed shoes	20	–	Muscular and heavy male. Post-mortem sharp-force breakage to an ankle	
Viminacium (Serbia) AD early 4th century	Cemetery: brick and tile grave	None	23–33	169–177	Enthesophytes to scapulae, pelvis, knees and ankles. Peri-mortem weapon injuries: projectile perforation to the right ilium; embedded iron arrowhead to the posterior aspect of the right femoral greater trochanter; weapon injury to the right parietal bone.	Golubović et al., 2009
Dura-Europos (Syria) AD 265	Brick wall within city, built to support defensive towers 25 & 26	Unknown: "A well preserved skeleton … In this case we probably have to do with the burial of a warrior who perished during the siege of the city"	–	-		Toll 1946: 6. Individual re-buried on site
	Tower 19: mine	Armour and weaponry	19 adult males	–	-	James 2011b. Individuals most likely to have been re-buried in back-fill post-excavation
Acre - Ptolemais, (Israel) Roman period	Cemetery: stone-lined cist grave	Tombstone: 'Olpius Martinus'	adult male	–	-	Tepper and Nagar 2013
	Cemetery: stone-lined cist grave	Coin (tetradrachma) hoard	adult male	–	-	Tepper and Nagar 2013
Kalkriese (Germany) AD 9	Multiple burial deposits: disarticulated human bone	Armour and weaponry	adult males	–	Peri-mortem weapon and blunt force injuries	Grosskopf 2007

The army contained many ethnically diverse auxiliary units, whose presence is attested through inscription evidence and material culture (Haynes, 2013, James, 2011a) (see 2.2). Epigraphic evidence from civilian and military tombstones has long-been recognised as being a very skewed and biased source of data, particularly the accuracy of an individual's age-at-death; therefore, demographic data from this source is excluded from this discussion (Parkin, 1992; Harlow and Laurence, 2002; Revell, 2005; Carroll, 2006).

The identification of adult males still wearing their military equipment provides us with an opportunity to create an osteobiography (Saul and Saul, 1989; Buikstra and Scott, 2009; Stodder and Palkovich, 2012) which could provide an analogy to understand patterns of disease and injury in other Roman cemeteries (Table 6.1). The enthesopathy and osteoarthritis data are summarised in Table 6.1, but were not explored any further here, because recent research has demonstrated that these are more likely to be age-related in origin rather than being created by occupational patterns (Weiss and Jurmain, 2007; Henderson and Alves Cardoso, 2013). As Table 6.1 shows, the number of secure contexts is very limited and many of the individuals have peri-mortem injuries which are most likely to be associated with the manner of their death, emphasised by the clandestine nature of their burial, such as their body being disposed of down a well (Constandse-Westermann, 1982). The limited bioarchaeological data suggest that the only common factor was their age, between 20 and 40 years old.

Creating an osteobiography of the people who formed the wider Roman military community is highly problematic, because of the diverse nature of those groups. For example, it could be argued that Rome was a military settlement, because of the permanent presence of the Praetorian Guard in the city (Rankov, 2001) and therefore, all of its population were in some way connected to the military. Nevertheless, working pragmatically, building on the conclusions reached by Andersen (2009) and the archaeological evidence for families, slaves and servants in military settlements (army and civilian) (see sections 1 and 2.1) it may be possible to investigate this topic by examining variables such as cemetery demography, stable isotope evidence for diet and mobility, individuals buried with equipment associated with the military, individuals buried with imported or unusual grave-goods, in a low-status grave, or cremated burials. It is acknowledged that this creates a large 'net' but in the northwest provinces where epigraphy was not wide-spread (Mattingly, 2011), other avenues must be explored.

3. Materials and Methods

3.1. *Londinium: Settlement Overview and Military Connections*

Londinium was established following the Claudian invasion of AD 43. It was founded by *c.* AD 48, the settlement was planned and was more civilian than military in nature (Hill and Rowsome, 2011) (Figure 6.1). By AD 60, *Londinium* was described by Tacitus as a centre of commerce and bustling with traders, but it was not a *colonia*. This is supported by the diverse nature of the material culture from this period, much imported from the near Continent and the southern and eastern Mediterranean. The initial settlement was short-lived, destroyed by the Boudican rebellion in AD 60, but was rebuilt under a new Procurator and quickly regained its prominence in commercial activities however, another devastating fire occurred between AD 125–130. Occupation appears to have gone through a series of peaks and troughs, reflecting wider political uncertainties in the Empire, particularly when Britain briefly broke from the Empire in AD 286 and when Imperial control was withdrawn finally in AD 410 (Merrifield, 1983; Marsden, 1986; Perring, 1991; Bird *et al.*, 1996; Millett, 1996).

Focusing on its military connections, the fort was continually occupied throughout the period and in the early 2nd century AD, a large, complex and high status building was constructed in Southwark and is believed to been used by the provincial government (Hall, 2010). Soldiers from the three legions stationed in Britain at Chester, Caerleon and York were sent to the city to work for the Governor, exemplified by the tombstone of a clerical soldier found at Chamomile Street (Figure 6.2) (Hall, 2010). A wealth of military equipment, ranging from tent pegs to armour has also been recovered, showing that both cavalry and legionary forces were stationed in the city (Hall, 2010). One of the two remarkable finds indicating their presence is a 1st century AD helmet which has the names of four owners inscribed, meaning that it had been in use for at least 100 years (The British Museum, 2013). The other is a funerary inscription, which recorded the name of *Lucius Pompeius Licetus* from Arretium in Italy (modern Arezzo, Tuscany), a town which supplied recruits to the Roman army from the 1st century AD onwards (Museum of London, 2013).

One of the most authoritative overviews of the cemeteries of Roman London has been published by Hall (1996). Both cremations and inhumations are attested from the 1st to the early 5th century AD. Focusing on inhumations, a wide-range of coffin types (wood, stone, lead), burial treatments (gypsum or chalk coatings), tombstones and a limited number of mausolea have been identified but, overall, very few high-status burials have been recovered (Hall, 1996). It should be noted that despite the strong civilian nature of the settlement, the majority of tombstones found in the city are military (Mattingly, 2011). A fairly standardised range of grave goods are also present: jewellery, hobnailed shoes, pottery, glass products, coins and animal bone; nevertheless, the burials of London seem to be less wealthy than elsewhere in the province (Hall, 1996). However, at the present time, apart from the eastern cemeteries of Roman London, the bioarchaeological and funerary data have not been synthesised at city-level, and so it remains difficult to place individuals in their wider context.

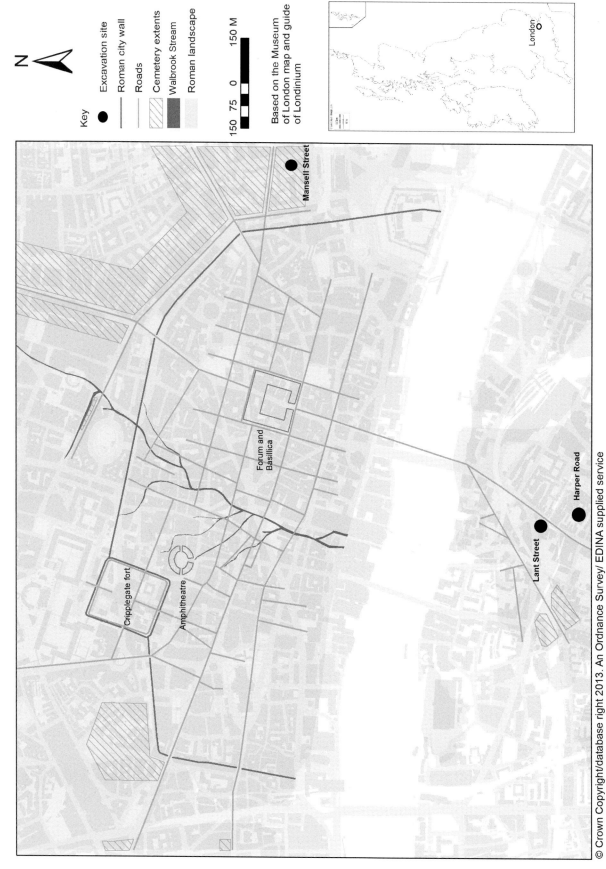

Figure 6.1 Location map of London, and map of Roman London showing the location of the three burials and other major public buildings in the settlement © Museum of London (adapted by Nathan Thomas).

Figure 6.2 The tombstone of a Roman soldier from Chamomile Street, London. The man is identified as a soldier because of his cloak, tunic, sword and hairstyle. He is carrying a scroll and six writing tablets in his left hand, indicating that he performed clerical duties for the military administration in Londinium © Museum of London.

Figure 6.3 The copper-alloy chip-carved belt set (two belt plates, buckle, and strap end) found in the AD 350–410 burial of an adult male from Mansell Street (© Museum of London).

3.2. Bioarchaeological Data

As discussed in section 2, creating an osteobiographical model for serving soldiers and/or the wider army community proved insurmountable – it could include everyone. Instead, a scaled-back and more achievable approach was taken, integrating funerary, stable isotope and bioarchaeological data.

Data for inhumation burials in the city of London (Figure 6.1) held by the London Archaeological Archive and Research Centre and the Centre for Human Bioarchaeology at the Museum of London (recorded using Connell and Rauxloh, 2006; Powers, 2012) were interrogated for the following variables: stable isotope evidence for diet and mobility, individuals buried with equipment associated with the military, individuals buried with imported or an unusual range of grave-goods. For the purposes of this paper, three burials were selected to explore the hypotheses described in section 1.

4. Results (Figure 6.1)

4.1. Men in the Military Community: 'German Man' (MSL87 sk 724)

This is a one of the late burials from Mansell Street, which forms part of the Eastern Roman cemetery, dating from AD 350 to 410 (Barber and Bowsher, 2000). The adult male is over 45 years old and was buried in a wooden coffin with two glass bottles, a copper alloy coin, a crossbow brooch was found on the right side of the torso and a belt buckle by the left arm (Barber and Bowsher, 2000) – these finds have resulted in him being colloquially known as 'German Man'. The belt buckle is chip-carved and appears to be unworn (Figure 6.3), further emphasized by the fact it is by his hand rather than on his torso and is one of three found in *Londinium* (Barber and Bowsher, 2000; Watson, 2014). The brooch and belt are very distinctive items in late Roman burials, crossbow brooches were used to fasten heavy outer garments at the shoulder, particularly cloaks, and are considered to have formed part of the uniform of a 4th century soldier or state official who had achieved a certain rank. The distribution of these brooch types is biased towards military zones but they have also been found in the burials of women and children (Collins, 2010). They are believed to indicate a high social status (of both sexes) and could perhaps suggest that the wearer spent a period of time in Roman service, such as a military officer (Collins, 2010, 2013). The belt buckle was synonymous with Roman soldiers, with primary sources remarking that it enabled them to be identified as a distinctive social group when not dressed in full-armour (Hoss, 2011, 2013). These belts were both practical and symbolic to soldiers; they were made at camps and issued by the army; their sword hung from it; it could be used instead of money; and could be confiscated as a form of punishment (Hoss, 2011). Like the brooch, the chip-carved style is considered to have Germanic military connections, and the wearing of belts by veterans may reflect an honourable discharge (Hoss,

Figure 6.4 Bronze torc with a feather and ring-and-dot decoration found in the 1st century AD burial of Harper Road Woman (© Museum of London).

2011; see also Booth, 2014). This and the crossbow brooch suggest personal, cultural or ancestral connections to Roman Germany and the military (Barber and Bowsher, 2000; Pearce, 2010), as his stable isotope results show that he is local to Britain (Shaw *et al.*, 2016).

The bioarchaeological analysis of his skeleton only found cribra orbitalia, and enamel hypoplastic defects and calculus to the dentition. Due to the high degree of bone fragmentation, it was not possible to estimate his stature.

4.2. Women in the Military Community: Harper Road Woman (HR79 sk 311)

This is one of the earliest inhumation burials in *Londinium*, dating from AD 50 to 70 and shows a hybridisation of late Iron Age and Roman identities. The

burial was discovered in modern-day Southwark, south of the River Thames, the opposite side of the river to the settlement founded in AD 48. In the Roman period, archaeological evidence suggests that this burial ground was on the outskirts of the settlement which had grown-up in a less organised fashion than that on the north bank (Cowan *et al.*, 2009). The burial contained the coffined skeleton of a female over 45 years old; her body was accompanied by both late Iron Age and Roman items, consisting of a flagon, a bronze mirror and a decorated bronze torc (Figure 6.4). The hybridised style of funeral can be traced in all of the items chosen for her burial. Coffins were used during the late Iron Age in Britain, but are also a feature of Romano-British inhumations (Whimster, 1981; Philpott, 1991). The flagon is a Roman

Figure 6.5 A folding iron-bladed knife with an inlaid ivory handle in the shape of a leopard probably consuming some meat, with an attached copper alloy chain, found in the 4th century AD burial of a subadult from Lant Street (© Pre-Construct Archaeology).

form and may have been made locally, but it could have been produced in France or Germany; the torc is the classic late Iron Age form of personal ornamentation (Champion, 1996), and appears to have been broken before burial – perhaps when it was removed from her neck? The mirror was made in north Italy and is a rare inclusion in female burials in Britain, both in the late Iron Age and Roman period (Barber and Bowsher, 2000; Joy, 2011). In the late Iron Age, mirrors played an important role in ritual practices and are believed by many scholars to be the female equivalent of male weapon burials (Giles and Joy, 2007). In the Roman period, mirrors are also believed to have been both practical and ritual items, and are considered to be high-status funerary goods (Addey, 2007).

Bioarchaeological analysis of her skeleton did not observe any diseases; only calculus and wear were observed on the dentition. Due to the high degree of bone fragmentation, it was not possible to estimate her stature. Analysis of the oxygen, strontium and lead isotopes from her dentition suggests that she had been exposed to anthropogeneic sources of lead, with the strontium and oxygen results indicating that she had not grown up in the London area, but rather in a location with Triassic to Lower Jurassic sedimentary rocks, 'at least 150km to the north and west of London' or the Continent (Budd, 2003).

4.3. Subadults in the Military Community: Lant Street Adolescent (LTU03 sk 385)

This burial was located in a cemetery established in the 4th century in modern-day Southwark, outside the settlement on the south bank of the River Thames (Sayer, 2005). The 14 year old female was laid on a bed of chalk and clay and had the richest burial in the cemetery (Sayer, 2005; Arthur, 2014). Their collection of grave-goods is unique in *Londinium*, on either side of their head was placed a glass oil flask (aryballos), and a spiral-trail vessel

in the shape of an amphora (amphorisk). At their feet the following objects were discovered (Figure 6.5): a folding iron-bladed knife with an inlaid ivory handle in the shape of a leopard probably consuming some meat – a copper alloy chain was attached to the knife; a lever lock key; and an ivory inlaid casket, with the ivory shaped into a female bust and numerous triangular and rectangular fragments, some decorated – this casket is considered to have been created as a funerary piece (Ridgeway, pers.comm; Sayer, 2005). The amphorisk has parallels to German finds but the only similar piece to the casket is a large ivory plaque depicting a male bust found at Wroxeter (Shropshire), a large settlement which began as a legionary fortress; and the knife is believed to have been made on the Continent (Sayer, 2005; Ridgeway, 2013). The burial as a whole is suggested as being Carthaginian (North African) in style and of high status (Ridgeway, 2013; Sayer, 2005). In a Roman life course perspective, if menarch had begun, they would have been considered 'adult' and could have been married to an older man (Harlow and Laurence, 2002)

The 14 year old has a number of congenital disorders to the axial and appendicular skeleton: bilateral calcaneal-navicular fibrous coalition, six lumbar vertebra are present and there are also separate transverse processes for the 12th thoracic vertebra and 6th lumbar vertebra. The radii and ulnae have changes consistent with residual rickets, and healing non-specific periostitis is present to the leg bones. A number of diseases and non-metric changes were recorded on the dentition: a talon cusp to the left 1st maxillary incisor, enamel hypoplastic defects are observable on the 1st maxillary and mandibular molars and calculus is present on the teeth.

Stable isotope data for mobility suggests that she had migrated from North Africa to *Londinium*, and the dietary results indicate that she had consumed a local diet for at least four to five years before her death (Millard *et al.*, in prep). Her journey to London and the symbolism of the grave goods included in her burial raise a number of potential interpretations. Despite the vast geographic distance between Britain and Africa, there were strong connections between the two, because after the Mauretanian war (Algeria) in the AD 140s, many North Africans had become auxillary troops or members of British legions, as British reinforcements had served there in the late 2nd century AD, and three governors of Britannia are known to have come from this region (Mattingly, 2006). It should be noted that epigraphic evidence from this region shows that serving soldiers had formal and 'common-law' marriages with local women and the daughters of military personnel stationed there (Raven, 1993; Cherry, 1998). In both *Londinium* and *Britannia*, North African people are also attested in inscription evidence on tombstones and pot graffiti (Mattingly, 2006), and by their distinctive pottery forms (see Swan, 1992). Consequently, it may well be that she was the daughter or wife of a military officer.

5. Discussion and Conclusions

Londinium was a cosmopolitan settlement which had a strong military presence, and the tombstone evidence suggests that we should expect to 'find' serving members of the military in its cemeteries. However, at the present time, its cemeteries lack secure osteological and funerary evidence for serving soldiers. The remains of Roman soldiers from elsewhere in the Empire provide an insight into a particular legionary or event (Table 6.1), but their use as an model for other locales and for burials of a different date, does not appear to be robust enough for contemporary research needs, especially in the light of the revised interpretations of activity markers (Henderson and Alves Cardoso, 2013). Taking into account the difficulties in identifying the graves of those who died in service (Andersen, 2009) and the realities of the Roman military, our attention should, instead, be drawn to the wider military community who may have expressed this relationship in their burial. It is recognised that a person's burial is created by their social group (amongst others, Bartel, 1982; Cannon, 1989) but in the Roman period, status and identity were also expressed in funerary treatment (Toynbee, 1996). We know that the Roman Empire, like its military, was not homogenous (Garnsey and Saller, 1987) and therefore, any attempt to identify members of this community must be spatially (and where possible, temporally) specific.

The burials described in section 4 all include elements which suggest a military connection, either their migration to Britain or the grave goods chosen to accompany them in death. These elements have links to military zones internal and external to *Britannia*, particularly the Rhineland and North Africa. Harper Road Woman's biological age and date of burial point to her being born before the Claudian invasion of Britain (AD 43), and therefore her presence in a Roman settlement indicates that she had reached *Londinium* from western or northern England or even the Continent, with her hybrid burial reflecting this transition period, with the settlement acting as a pull for migrants. How and why she came to *Londinium* remains open to speculation, and it is possible that she was enslaved. Favoured slaves could receive a higher status burial, but the type and variety of grave-goods does not suggest that she was enslaved (Joshel, 2010; George, 2013). The well-attested movement of the military community and their families between *Britannia* and North Africa may explain how the Lant Street adolescent came to be in *Londinium*, a hypothesis which is supported by the connections of the grave goods to military zones elsewhere in the province and Continent. The high status of these goods, similarly those of Harper Roman Woman, again does not suggest that they had been enslaved but this does remain a possibility - the absence of epigraphic evidence to further explore how they came to be in *Londinium* hinders our ability to explore these issues with greater certainty (Joshel, 2010; Galvao-Sobrinho, 2012). The man from Mansell Street has perhaps the strongest and most direct connections to the Roman military through his grave-goods, particularly the belt pieces (Collins, 2010, 2013; Hoss, 2013). As he was local to Britain, the type and placement of the objects in his grave, hints rather at cultural and/or ancestral connections to Roman Germany and the army (Barber and Bowsher, 2000; Pearce, 2010), and may rather reflect his identity as a veteran or soldier. However, Gardner's (2007) discussion of the military in late Roman Britain cautions against us expecting to see a specific group of Roman 'official' items and reminds us that 'soldier' or 'official' was only one part of a person's identity.

Our ability to identify different social groups in the Roman world remains highly problematic, even when epigraphic evidence is present. Bioarchaeological data can play a major role in interpretation, because the evidence for identity captured by the human body provides an insight into a person's life course and experiences that cannot be retrieved from any other source (Gowland and Thompson, 2013). These case-studies have shown that our understanding of identity based on grave-goods alone is not robust enough to explore the complex nature of lives in the Roman Empire (e.g. Eckardt *et al.*, 2009, 2014; Cool, 2010). The valuable contribution that osteobiographies can make is acknowledged but perhaps is more suited to burials where epigraphy is extant, or those with secure military contexts, such as the soldier buried down a well at Velsen (Table 6.1). Additionally, it is also considered that creating a model from 'known' soldiers/veterans to use in the absence of a distinctive burial pattern (Andersen, 2009) no longer meets the needs of contemporary methodological approaches. Therefore, in Roman London, it is posited that at the current time is it not possible to identify military personnel who died in service. By widening the scope to the military community, and undertaking a holistic approach with the available evidence from this settlement, it may be possible to explore the possibility that many of the people buried in its cemeteries had military connections that, in the absence of tombstones, are expressed through their diet, origin and grave goods.

Acknowledgements

I am grateful to the work of previous members of the Centre for Human Bioarchaeology for contributing so much to the WORD project, especially Richard Mikulski for his work on the Roman cemeteries. Thanks are also given to Caroline MacDonald, Roy Stephenson and Jelena Bekvalac at MoL, and Victoria Ridgeway from Pre-Construct Archaeology.

The stable isotope results included in this research could not have been achieved without the collaborative work and funding of Rebecca Gowland, Andrew Millard and Darren Gröcke, and their students, Lucie Johnson and Lindsay Powell (Durham University). Express thanks are given to Simon James, John Pearce and Jenny Hall; also Stefanie Hoss and Rob Collins for their advice and literature on the belt and brooch buried with 'German Man'; and Nathan Thomas for creating the map.

References

Addey, C. 2007. Mirrors and divination: Catoptromancy, oracles and earth goddesses, in: Andersen, M., (Ed.), *The Book of The Mirror: An Interdisciplinary Collection Exploring The Cultural Story of The Mirror.* Cambridge Scholars Publishing, Newcastle upon Tyne, pp. 32–46.

Allison, P.M., Fairbairn, A.S., Ellis, S.J.R., Blackall, C.W. 2005. Extracting the social relevance of artefact distribution in Roman military forts. *Internet Archaeology,* 17, doi 10.11141/ia.17.4. https://intarch.ac.uk/journal/issue17/allison_index.html (accessed 17th October, 2013).

Andersen, L.M. 2009. *The Roman military community as expressed in its burial customs during the first to third centuries CE.* PhD thesis, Brown University. http://library.brown.edu/find/Record/bdr102 (accessed 13th April, 2013).

Arthur, N. 2014. *Coming of age: the timing of puberty in Roman Britain, accessed through newly-developed osteological methods.* Unpublished MSc Thesis, Durham University.

Barber, B., Bowsher, D. (Eds.), 2000. *The Eastern Cemetery of Roman London. Excavations 1983–1990.* Museum of London Archaeology Monograph 4, London.

Bartel, B. 1982. A historical review of ethnological and archaeological analyses of mortuary practice. *Journal of Anthropological Archaeolaeology* 1, 32–58.

Bennett, D., Frere, S.S., Stow, S. (Eds.), 1982. *Excavations at Canterbury Castle. Volume 1.* Canterbury Archaeological Trust Ltd and Kent Archaeological Society, Canterbury.

Bird, J., Hassall, M., Sheldon, H. (Eds.), 1996. *Interpreting Roman London. Papers in Memory of Hugh Chapman.* Oxbow Books, Oxford.

Bishop, M.C. 1990. On parade: Status, display, and morale in the Roman Army, in: Busch, A.W., Schalles, H-J., (Eds.), *Akten der 14. Internationalen Limeskongresses in Bad Deutsch-Altenburg/Carnuntum, 14.-21. September, 1986, Römische Limes in Österreich Sonderband, (Vienna 1990).* Verlag Philipp von Zabern, Mainz, pp. 21–30.

Bishop, M.C. 2011. Weaponry and military equipment, in: Allason-Jones, L., (Ed.), *Artefacts in Roman Britain: Their Purpose and Use.* Cambridge University Press, Cambridge, pp. 114–132.

Bodel, J. (Ed.), 2001. *Epigraphic Evidence. Ancient History from Inscriptions.* Routledge, London.

Booth, P. 2014. A late Roman military burial from the Dyke Hills, Dorchester on Thames, Oxfordshire. *Britannia* 45, 243–273.

Bowman, A.K. 2004. *Life and Letters on the Roman Frontier. Vindolanda and Its People.* British Museum Press, London.

Budd, P. 2003. Combined O-, Sr- and Pb- isotope analysis of dental tissues from a Neolithic individual from Shepperton and an Iron Age individual from Southwark, London. Archaeotrace Ltd Report No. 106.

Buikstra, J.E., Beck, L.A. (Eds.), 2006. *Bioarchaeology: The Contextual Analysis of Human Remains.* Academic Press, London.

Buikstra, J.E., Scott, R.E. 2009. Key concepts in identity studies, in: Knudson, K.J., Stojanowski, C.M. (Eds.), *Bioarchaeology and Identity in the Americas.* University of Florida Press, Gainsville, FL, pp. 24–55.

Caesar, Julius. (Trans. Handford, S.A.) 1983. *The Conquest of Gaul.* Penguin Classics, London.

Cannon, A. 1989. The historical dimension in mortuary expressions of status and sentiment. *Current Anthropology* 30, 437–458.

Carroll, M. 2006. *Spirits of the Dead: Roman Funerary Commemoration in Western Europe,* Oxford University Press, Oxford.

Carroll, M. 2012. Vindonissa (Brugg, Switzerland), in: Bagnall, R.S., Brodersen, K., Champion, C.B., Erskine, A., Huebner, S.R. (Eds.), *The Encyclopedia of Ancient History.* Wiley-Blackwell, Oxford. http://onlinelibrary.wiley.com/book/10.1002/9781444338386 (accessed 15th August, 2013).

Champion, S. 1996. Jewellery and adornment, in: Green, M. (Ed.), *The Celtic World.* Routledge, London.

Chenery, C., Eckardt, H. and Müldner, G. 2011. Cosmopolitian Catterick? Isotopic evidence for population mobility on Rome's northern frontier. *Journal of Archaeological Science* 38, 1395–1770.

Cherry, D. 1998. *Frontier and Society in Roman North Africa.* Clarendon Press, Oxford.

Cheung, C., Schroeder, H., Hedges, R.E.M. 2012. Diet, social differentiation and cultural change in Roman Britain: New isotopic evidence from Gloucestershire. *Archaeological and Anthropological Sciences* 4, 61–73.

Collins, R. 2010. Brooch use in the 4th-to 5th-century frontier, in: Collins, R., Allason-Jones, L., (Eds.), *Finds from the Frontier: Material Culture in the 4th–5th Centuries.* Council for British Archaeology, York, pp. 64–77.

Collins, R. 2013. Soldiers to warriors: Renegotiating the Roman frontier in the fifth century, in: Hunter, F., Painter, K. (Eds.), *Late Roman Silver: The Traprain Treasure in Context.* Society of Antiquaries of Scotland, Edinburgh, pp. 29–43.

Connell, B., Rauxloh, P. 2006. *A Rapid Method for Recording Human Skeletal Data.* Museum of London, London. http://archive.museumoflondon.org.uk/Centre-for-Human-Bioarchaeology/Database (accessed 29th November, 2013).

Constandse-Westermann, T.S. 1982. A skeleton found in a Roman well at Velsen (Province North Holland. The Netherlands). *Helinium* XXII, 135–167.

Cool, H.E.M. (Ed.), 2004. *The Roman Cemetery at Brougham, Cumbria: Excavations 1966–67.* Britannia Monograph Series 21, London.

Cool, H.E.M. 2006. *Eating and Drinking in Roman Britain.* Cambridge University Press, Cambridge.

Cool, H.E.M. 2010. Finding the foreigners, in: Eckardt, H., (Ed.), *Roman Diasporas: Archaeological Approaches to Mobility and Diversity in the Roman Empire.* Journal of Roman Archaeology Supplementary Series 78, pp. 27–44.

Cowan, C., Seeley, F., Wardle, A., Westman, A., Wheeler, L. (Eds.), 2009. *Roman Southwark Settlement and Economy: Excavations in Southwark 1973–91.* Museum of London Archaeology Monograph 42, London.

Craig, O.E., Biazzo, M., O'Connell, T.C., Garnsey, P., Martinez-Labarga, C., Lelli, R., Salvadei, L., Tartaglia, G., Nava, A., Reno, L., Fiammenghi, A., Rickards, O., Bondioli, L. 2009. Stable isotopic evidence for diet at the Imperial Roman coastal site of Velia (1st and 2nd centuries AD) in southern Italy. *American Journal of Physical Anthropology* 139, 572–583.

Croom, A.T., Caffell, A. 2005. Human remains from South Shields Roman Fort and its cemetery. *The Arbeia Journal* 8, 101–116.

Cucina, A., Vargiu, R., Mancinelli, D., Ricci, R., Santandrea, E., Catalano, P., Coppa, A. 2006. The necropolis of Vallerano (Rome, 2nd–3rd century AD): An anthropological perspective on the ancient Romans in the Suburbium. *International Journal of Osteoarchaeology* 16, 104–117.

Cummings, C. 2008. *Food and society in late Roman Britain: determining dietary patterns using stable isotope analysis.* Unpublished PhD Thesis, University of Oxford.

Davies, D.W. 1989. *Service in the Roman Army.* Edinburgh University Press, with the Publications Board of the University of Durham, Edinburgh.

Davis, J.B., Thurnam, J. 1865. *Crania Britannica: Delinations and Descriptions of the Skulls of the Aboriginal and Early Inhabitants of the British Islands with Notices of Their Other Remains.* Taylor & Francis, London.

Dupras, T., Schwartz, H.P., Fairgrieve, S.I. 2001. Infant feeding and weaning practices in Roman Egypt. *American Journal of Physical Anthropology* 115, 204–212.

Eckardt, H., (Ed.), 2010. *Roman diasporas: Archaeological approaches to mobility and diversity in the Roman Empire.* Journal of Roman Archaeology Supplementary Series 78.

Eckardt, H., Booth, P., Chenery, C., Müldner, G., Evans, J.A., Lamb, A. 2009. Isotopic evidence for mobility at the late Roman cemetery at Lankhills, Winchester. *Journal of Archaeological Science* 36, 2816–2825.

Eckardt, H., Chenery, C., Leach, S., Lewis, M., Müldner, G., Nimmo, E. 2010. A long way from home: Diaspora communities in Roman Britain, in: Eckardt, H., (Ed.), *Roman Diasporas: Archaeological Approaches to Mobility and Diversity in the Roman Empire.* Journal of Roman Archaeology Supplementary Series 78, pp. 99–130.

Eckardt, H., Müldner, G., Lewis, M. 2014. People on the move in Roman Britain. *World Archaeology* 46.4, 534–550.

Erdkamp, P. 2007. *A Companion to the Roman Army.* Oxford University Press, Oxford.

Evans, J., Stoodley, N., Chenery, C. 2006. A strontium and oxygen isotope assessment of a possible fourth century immigrant population in a Hampshire cemetery, southern England. *Journal of Archaeological Science* 33, 365–372.

Fowler, C. 2004. *The Archaeology of Personhood. An Anthropological Approach.* Routledge, London.

Galvao-Sobrinho, C.R. 2012. Feasting the dead together: Household burials and the social strategies of slaves and freed persons in the early Prinicipate, in: Ramsby, T., Bell, S. (Eds.), *Free At Last! The Impact of Freed Slaves on the Roman Empire.* Bloomsbury, London, pp. 130–176.

Gardner, A. 2007. *An Archaeology of Identity. Soldiers and Society in Late Roman Britain.* Left Coast Press, California.

Garnsey, P., Saller, P. 1987. *The Roman Empire: Economy, Society and Culture.* University of California Press, California.

George, M. 2013. Introduction, in: George, M. (Ed.), *Roman Slavery and Roman Material culture.* University of Toronto Press, London, pp. 3–18.

Giles, M., Joy, J. 2007. Mirrors in the British Iron Age: Performance, revelation and power, in: Anderson, M. (Ed.), *The Book of the Mirror. An Interdisciplinary Collection Exploring the Cultural Story of the Mirror.* Cambridge Scholars Press, Newcastle upon Tyne, pp. 16–31.

Goldsworthy, A., Haynes, I. (Eds.), 1999. *The Roman army as a community. Including papers of a conference held at Birkbeck College, University of London on 11–12 January, 1997.* Journal of Roman Archaeology Supplementary Series 34.

Golubović, S., Mrdjić, N., Speal, C.S. 2009. Killed by the arrow: Grave No. 152 from Viminacium, in: Busch, A.W., Schalles, H.-J. (Eds.), *Waffen in Aktion. Akten der 16 Internationalen Roman Military Equipment Conference (ROMEC). Xanten, 13–16 Juni 2007.* Philipp von Zabern, Mainz, pp. 55–64.

Gowland, R. 2001. Playing dead: Implications of mortuary evidence for the social construction of childhood in Roman Britain, in: Davies, G., Gardner, A., Lockyear, K. (Eds.), *TRAC 2000. Proceedings of the Tenth Annual Theoretical Roman Archaeology Conference, London 2000.* Oxbow Books, Oxford, pp. 152–168.

Gowland, R. 2006. Ageing the past: Examining age identity from funerary evidence, in: Gowland, R., Knüsel, C.J. (Eds.), *Social Archaeology of Funerary Remains.* Oxbow Books, Oxford, pp. 143–154.

Gowland, R.L., Chamberlain, A.T. 2002. A Bayesian approach to ageing perinatal skeletal material from archaeological sites: implications for the evidence for infanticide in Roman Britain. *Journal of Archaeological Science* 29, 677–685.

Gowland, R., Thompson, T. 2013. *Human Identity and Identification.* Cambridge University Press, Cambridge.

Hall, J. 1996. The cemeteries of Roman London, in: Bird, J., Hassall, M., Sheldon, H. (Eds.), *Interpreting Roman London. Papers in Memory of Hugh Chapman.* Oxbow Books, Oxford, pp. 57–84.

Hall, J. 2010. *Londinium Lite.* Museum of London. http://archive.museumoflondon.org.uk/Londinium/ (accessed 21st October, 2013).

Harlow, M., Laurence, R. 2002. *Growing Up and Growing Old in Ancient Rome: A Life Course Approach.* Routledge, London.

Haynes, I. 2013. *Blood of the Provinces. The Roman Auxilia and the Making of Provincial Society from Augustus to the Severans.* Oxford University Press, Oxford.

Henderson, C.Y., Alves Cardoso, F. 2013. Special issue entheseal changes and occupation: Technical and theoretical advances and their applications. *International Journal of Osteoarchaeology* 23, 127–134.

Hill, J., Rowsome, P. 2011. *Roman London and the Walbrook Stream Crossing: Excavations at 1 Poultry and Vicinity, City of London.* Museum of London, London.

Hingley, R. 2000. *Roman Officers and English Gentlemen.* Routledge, London.

Hingley, R. 2005. *Globalizing Roman Culture. Unity, Diversity and Empire.* Routledge, London.

Hingley, R. 2009. Romanization, Haverfield and Imperial Britain. *Journal of Roman Archaeology,* 22, 705–709.

Hope, V.M. 1997a. Constructing Roman identity: Funerary monuments and social structure in the Roman world. *Mortality* 2, 103–121.

Hope, V.M. 1997b. Words and pictures: The interpretation of Romano-British tombstones. *Britannia* 28, 245–258.

Hope, V.M. 2003. Trophies and tombstones: Commemorating the Roman soldier. *World Archaeology* 35, 79–97.

Hoss, S. 2011. The Roman military belt, in: Koefoed, H., Nosch, M.-L. (Eds.), *Wearing the Cloak. Dressing the Soldier in Roman Times. Ancient Textile Series* 10, Oxbow Books, Oxford, pp. 29–44.

Hoss, S. 2013. A theoretical approach to Roman military belts, in: Sanader, M., Rendić-Miočević, A., Tončinić, D., Radman-Livaja, I. (Eds.), *Proceedings of the XVIIth Roman Military Equipment Conference: Weapons and Military Equipment in a Funerary Context (XVII Roman Military Equipment Conference, Zagreb, 24th-27th May, 2010).* ROMEC, Zagreb, pp. 317–26.

Insoll, T. (Ed.), 2007. *The Archaeology of Identities, A Reader.* Routledge, London.

James, S. 1999. The community of soldiers: A major identity and centre of power in the Roman Empire, in: Baker, P., Forcey, C., Jundi, S., Witcher, R. (Eds.) *TRAC 98: Proceedings of the Eighth Annual Theoretical Roman Archaeology Conference.* Oxbow Books, Oxford, pp. 14–25.

James, S. 2002. Writing the Legions: The development and future of Roman military studies in Britain. *Archaeological Journal* 159, 1–58.

James, S. 2006. Engendering change in our understanding of the structure of Roman military communities. *Archaeological Dialogues* 13, 31–36.

James, S. 2011a. *Rome and the Sword: How Warriors and Weapons Shaped Roman History.* Thames & Hudson Ltd, London.

James, S. 2011b. Stratagems, combat, and 'chemical warfare' in the siege mines of Dura-Europos. *American Journal of Archaeology* 115, 69–101.

Jones, R. 1984. Death and distinction in military and civilian Roman Britain, in: Blagg, T.F.C., King, A.C. (Eds.), *Military and Civilian in Roman Britain: Cultural Relationships in a Frontier Province.* British Archaeological Reports, British Series 136, Oxford, pp. 219–226.

Josephus. (Trans. Williamson, G.) 1981. *The Jewish War.* Penguin Classics, London.

Joshel, S.R. 2010. *Slavery in the Roman world.* Cambridge University Press, Cambridge.

Joy, J. 2011. Exploring status and identity in later Iron Age Britain: Reinterpreting mirror burials, in: Moore, T., Armada, X.L. (Eds.), *Atlantic Europe in the First Millennium BC: Crossing the Divide.* Oxford University Press, Oxford, pp. 468–487.

Juvenal. (Trans. Rudd, N.) 2008. *The Satires.* Oxford University Press, Oxford.

Keppie, L.J.F. 2000. *Legions and Veterans: Roman Army Papers 1971–2000.* Die Deutsche Bibliothek-CIP-Einheitsaufnahme, Stuttgart.

Killgrove, K. 2013. Biohistory of the Roman Republic: The potential of isotope analysis of human skeletal remains. *Post-Classical Archaeologies* 3, 41–62.

King, A.C. 1999. Animals and the Roman Army: Evidence of animal bones. *Journal of Roman Archaeology Supplementary Series* 34, 139–150.

Knapp, R. 2011. *Invisible Romans.* Harvard University Press, Cambridge, MA.

Knudson, K.J., Stojanowski, C.M. 2008. New directions in bioarchaeology: Recent contributions to the study of human social identities. *Journal of Archaeological Research* 16, 397–432.

Knudson, K.J., Stojanowski, C.M. (Eds.), 2009. *Bioarchaeology of Identity in the Americas.* University of Florida Press, Gainsville, FL.

Leach, S., Eckardt, H., Chenery, C., Müldner, G., Lewis, M. 2010. A 'lady' of York: migration, ethnicity and identity in Roman York. *Antiquity* 84, 131–145.

MacMullen, R. 1984. The Legion as a society. *Historia* 33, 440–456.

Mann, J.C. 2002. The settlement of veterans discharged from auxiliary units stationed in Britain. *Britannia* 33, 183–88.

Marsden, P. 1986. *Roman London.* Thames & Hudson, London.

Mattingly, D. 2006. *An Imperial possession: Britain in the Roman Empire, 54BC–AD409.* Penguin, London.

Mattingly, D. 2011. Urbanism, epigraphy and identity in the towns of Britain under Roman rule, in: Schellenberg, H.M., Hirschmann, V.E., Krieckhaus, A. (Eds.), *A Roman Miscellany: Essays in Honour of Anthony R. Birley on his seventieth birthday.* Akanthina Monograph Series 3, pp. 53–71.

McKinley, J. 2004. The human remains and aspects of pyre technology and cremation rituals, in: Cool, H.E.M. (Ed.), *The Roman Cemetery at Brougham, Cumbria: Excavations 1966–67.* Britannia Monograph Series 21, pp. 283–310.

Merrifield, R. 1983. *London. City of the Romans.* BT Batsford Ltd, London.

Millett, M. 1996. Characterizing Roman London, in: Bird, J., Hassall, M., Sheldon, H., (Eds.), *Interpreting Roman London. Papers in memory of Hugh Chapman.* Oxbow Books, Oxford, pp. 7–12.

Mrav, Z. 2013. Graves of auxiliary soldiers and veterans from the first century AD in the northern part of Pannonia, in: Sanader, M., Rendić-Miočević, A., Tončinić, D., Radman-Livaja, I. (Eds.), *Proceedings of the XVIIth Roman Military Equipment Conference: Weapons and Military Equipment in a Funerary Context (XVII Roman Military Equipment Conference, Zagreb, 24th-27th May, 2010).* ROMEC, Zagreb, pp. 87–116.

Müldner, G. 2005. *York: A Diachronic Study of Human Diet in York by Stable Isotope Analysis.* Unpublished PhD Thesis, University of Bradford.

Müldner, G. 2013. Stable isotopes and diet: Their contribution to Romano-British research. *Antiquity* 87, 137–149.

Museum of London, 2013. *Tombstone (HOO88[533]<75>* [Online]. Museum of London. http://archive.museumoflondon. org.uk/RWWC/objects/record.htm?type=object&id=283552 (accessed 21st October, 2013).

Parkin, T.G. 1992. *Demography and Roman Society.* The John Hopkins University Press, London.

Pearce, J. 2008. Burial evidence from Roman Britain. The un-numbered dead, in: Scheid, J., (Ed.), *Pour une archaeologie du rite. Nouvelles perspectives de l'archaeologie funeraire. Etudes reunies par John Scheid.* Collection de L'Ecole Francaise de Rome, Paris, pp. 29–42.

Pearce, J. 2010. Burial, identity and migrations in the Roman world, in: Eckardt, H. (Ed.), *Roman Diasporas. Archaeological approaches to mobility and diversity in the Roman Empire.* Journal of Roman Archaeology Supplementary Series 78, pp. 79–98.

Perring, D. 1991. *Roman London.* Seaby, London.

Phang, S.E. 2008. *Roman Military Service. Ideologies of Discipline in the Late Republic and Early Prinicipate.* Cambridge University Press, Cambridge.

Philpott, R. 1991. *Burial Practices in Roman Britain: A survey of grave treatment and furnishing AD 43–410.* British Archaeological Reports, British Series 219, Oxford.

Pollard, A.M. 2011. Isotopes and impact: A cautionary tale. *Antiquity* 85, 631–638.

Powell, L.A., Redfern, R.C., Millard, A.R. 2014. Infant feeding practices in Roman London: the isotopic evidence, in: Carroll, P.M., Graham, E.-J. (Eds.), *Infant health and death in Roman Italy and beyond.* Journal of Roman Archaeology Supplementary Series 96, pp. 89–110.

Powers, N. (Ed.), 2012. *Human Osteology Method Statement.* Museum of London, London. http://archive.museumoflondon. org.uk/Centre-for-Human-Bioarchaeology/Database (accessed 29th November, 2013).

Prowse, T.L., Barta, J.L., Von Hunnius, T.E., Small, A.M. 2010. Stable isotope and mtDNA evidence for geographic origins at the site of Vagnari, South Italy, in: Eckardt, H. (Ed.) *Diasporas in the Roman world.* Journal of Roman Archaeology Supplementary Series 78, pp. 175–197.

Prowse, T.L., Schwarcz, H.P., Garnsey, P., Bondioli, L., Macchiarelli, R. 2004. Isotopic paleodiet studies of skeletons from the Imperial Roman-age cemetery of Isola Sacra. *Journal of Archaeological Science* 31, 259–272.

Prowse, T.L., Schwarcz, H.P., Garnsey, P., Knyf, M., Macchiarelli, R., Bondioli, L. 2007. Isotopic evidence for age-related immigration to Imperial Rome. *American Journal of Physical Anthropology* 132, 510–519.

Rankov, B. 2001. *The Praetorian Guard.* Osprey Publishing Ltd, Lancashire.

Raven, S. 1993. *Rome in Africa*, third ed. BT Batsford Ltd., London.

Redfern, R.C., DeWitte, S. 2011. A new approach to the study of Romanization in Britain: A regional perspective of cultural change in late Iron Age and Roman Dorset using the Siler and Gompertz-Makeham models of mortality. *American Journal of Physical Anthropology* 144, 269–285.

Redfern, R.C., Gowland, R.L. 2011. A bioarchaeological perspective on the pre-adult stages of the life course: Implications for the care and health of children in the Roman Empire, in: Harlow, M.L., Lovén, L.L. (Eds.), *Families in the Roman and Late Antique Roman World.* Continuum International Publishing Group, London, pp. 111–140.

Redfern, R.C., Hamlin, C., Beavan Athfield, N. 2010. Temporal changes in diet: A stable isotope analysis of late Iron Age and Roman Dorset, Britain. *Journal of Archaeological Science* 37, 1149–1160.

Redfern, R.C., Hamlin, C., Millard, A. 2012. A regional investigation of subadult dietary patterns and health in late Iron Age and Roman Dorset, England. *Journal of Archaeological Science* 39, 1249–1259.

Revell, L. 2005. The Roman life course: a view from the inscriptions. *European Journal of Archaeology* 8, 43–63.

Ridgeway, V., Leary, K., Sudds, B. 2013. *Roman Burials in Southwark. Excavations at 52-56 Lant Street and 56 Southwark Bridge Road, London SE1.* Pre-Construct Archaeology Limited Monograph No. 17, London.

Roth, J., Roth, J.P. 1998. *The Logistics of the Roman Army at War: 264 B.C.– A.D. 235.* Die Deutsche Bibliothek-CIP-Einheitsaufnahme, Netherlands.

Rowland, R.J. 1976. Foreigners in Roman Britain. *Acta Archaeologica Academiae Scientiarum Hungaricae* 28, 443–447.

Roxan, M.M. 1991. Women of the frontiers, in: Maxfield, V.A., Dobson, M. (Eds.), *Roman Frontier Studies 1989: Proceedings of the XVth International Congress of Roman Frontier Studies.* Exeter University Press, Exeter, pp. 462–467.

Roxan, M., Holder, P. 2003. *Roman Military Diplomas IV. BICS Supplement 82.* Institution of Classical Studies, London.

Saller, R.P., Shaw, B.D. 1984. Tombstones and Roman family relations in the Principate: civilians, soldiers and slaves. *The Journal of Roman Studies* 74, 124–156.

Saul, F.P., Saul, J.M. 1989. Osteobiography: A Maya example, in: İşcan, M. Y. and Kennedy, K.A.R. (Eds.), *Reconstruction of Life from the Skeleton.* Alan R Liss, New York, pp. 287–302.

Sayer, K. 2005. *An Assessment of an Archaeological Excavation at 52-56 Lant Street, London Borough of Southwark, LTU03.* Pre-Construct Archaeology Ltd, London.

Scheidel, W. 1996. Measuring sex, age and death in the Roman empire: Explorations in ancient demography. *Journal of Roman Archaeology Supplement Volume* 21.

Scheidel, W. 1997. Quantifying the sources of slaves in the early Roman Empire. *The Journal of Roman Studies* 87, 156–169.

Scheidel, W. 2005. Marriage, families, and survival in the Roman Imperial army: Demographic aspects. *Princeton/Stanford Working Papers in Classics.* http://www.princeton.edu/~pswpc/pdfs/scheidel/110509.pdf (accessed 23rd December, 2011).

Scheidel, W. 2010. Physical wellbeing in the Roman world. *Princeton/Stanford Working Papers in Classics.* http://www.princeton.edu/~pswpc/pdfs/scheidel/011001.pdf (accessed 22nd December, 2011).

Schmidt, C.W., Symes, S.A. (Eds.), 2008. *The Analysis of Burned Human Remains.* Academic Press, London.

Shaw, H., Montgomery, J., Evans, J., Redfern, R.C., Gowland, R. 2016. Identifying migrants in Roman London using lead and strontium stable isotopes. *Journal of Archaeological Science* 66, 57–68.

Sofaer, J.R. 2006. *The Body as Material Culture. A Theoretical Osteoarchaeology.* Cambridge University Press, Cambridge.

Stodder, A.L.W., Palkovich, A.M. (Eds.), 2012. *The Bioarchaeology of Individuals.* University of Florida Press, Gainsville, FL.

Swan, V.G. 1992. Legio VI and its men: African legionaries in Britain. *J Rom Pot Stud* 5, 1–33.

Swift, E. 2009. *Regionality in Dress Accessories in the Late Roman West.* Editions Monique Mergoil, Montagnac.

The British Museum, 2013. *Roman Legionary Helmet (1950, 0706.1).* The British Museum, London. http://www.britishmuseum.org/research/collection_online/collection_object_details.aspx?objectId=1383869&partId=1&searchText=roman+helmet,+London&page=1 (accessed 21st October, 2013).

Toynbee, J.M.C. 1996. *Death and Burial in the Roman World.* Thames & Hudson, London.

Vegetius F.R. 1993. *Epitome of Military Science.* Liverpool University Press, Liverpool (Trans. Milner, N.P.).

Watson, S. 2014. *The Pompeii of the North: Excavations at Bloomberg London and the Return of the Temple of Mithras.* Gresham College Lecture. https://www.academia.edu/8721556/Pompeii_of_the_North_Excavations_at_Bloomberg_London_and_the_Return_of_the_Temple_of_Mithras (accessed 6th October, 2014).

Webster, G. 1998. *The Roman Imperial Army of the First and Second Centuries A.D..* University of Oklahoma Press, Oklahoma.

Webster, J. 2005. Archaeologies of slavery and servitude: Bringing 'New World' perspectives to Roman Britain. *Journal of Roman Archaeology* 18, 161–179.

Webster, J. 2010. Routes to slavery in the Roman world: A comparative perspective on the archaeology of forced migration, in: Eckardt, H. (Ed.), *Roman diasporas. Archaeological approaches to mobility and diversity in the Roman Empire.* Journal of Roman Archaeology Supplementary Series 78, pp. 45–65.

Weiss, E., Jurmain, R. 2007. Osteoarthritis revisited: A contemporary review of aetiology. *International Journal of Osteoarchaeology* 17, 437–450.

Wells, L.H. 1957–8. The 'dwarf' skeleton from the Roman fort at Newstead. *Proc. R. Soc.* 139–143.

Wells, P.S. 2004. *The Battle That Stopped Rome.* W.W. Norton & Company, London.

Whimster, R. 1981. *Burial practices in Iron Age Britain. A discussion and gazetteer of the evidence c.700B.c.–A.D. 43.* British Archaeological Reports, British Series 99, Oxford.

Wilmott, T., Cool, H., Evans, J. (Eds.), 2009. *Excavations at the Hadrian's Wall Fort of Birdoswald (Banna), Cumbria: 1986–2000.* English Heritage, London.

7

3D Recording of Normal Entheses: A Pilot Study

Charlotte Y. Henderson

The study of past activity-patterns in skeletal remains is a widespread question, particularly using the entheses; zones at which the soft and hard tissues of the musculoskeletal system interface. These zones are complex and it is only recently, with the advent of cheaper equipment, that they have been recorded in three-dimensions. This pilot project aims to test a faster method for recording entheses in three-dimensions to calculate area and the root mean square (RMS) through a section of the enthesis at midpoint. The secondary aim was to determine variation in normal enthesis area and RMS and how this relates to proxies for body size. Identified skeletons of labourers curated in the identified skeletal collection in Coimbra were recorded using a NextEngine(TM) HD Desktop 3D laser scanner and its proprietary software. Results demonstrate that error, when data are collected over a long period of time, is a significant problem. No statistically significant associations were found between enthesis area and body size proxies nor for RMS and enthesis area. This contrasts with other three-dimensional studies and could be the effect of only recording normal entheses, the occupational category, small sample size or the error rates.

Keywords Biceps brachii; Common extensor origin; Coimbra identified skeletal collection; Labourers; Trabalhadores

1. Introduction

Entheses of the upper limb, in particular, have been widely used to study activity-patterns in the past (for reviews see Peterson and Hawkey, 1998; Jurmain, *et al.* 2012; Henderson and Alves Cardoso, 2013). Their intermediate role between the soft and hard tissues of the musculoskeletal system has led to the interpretation of changes at these sites, for example new bone formation, as indicators of repetitive movement (e.g. Hawkey and Merbs, 1995). The association of entheseal changes (EC) with the ageing process has, more recently, led researchers to encourage more limited interpretations (Alves Cardoso and Henderson, 2010, 2013; Milella *et al.*, 2012; Villotte and Knüsel, 2013). Nevertheless, their lure to those wishing to reconstruct activity-patterns in the past continues, as evidenced by the number of publications this author found during 2011 and 2012 (Deveci, 2011; Godde, 2011; Lieverse, 2011; Mariotti and Belcastro, 2011; Myszka and Piontek, 2011; Rojas-Sepúlveda, 2011; Santos *et al.*, 2011; Stefanović and Porčić, 2013;

Foster *et al.*, 2012; Humphries and Sarasúa, 2012; Jurmain *et al.*, 2012; Milella *et al.*, 2012; Myszka, Piontek and Niinimäki, 2012; Schlecht, 2012a, 2012b; Üstündağ and Shuler *et al.*, 2012; Villotte and Knüsel, 2012), a period chosen because it reflects publication rates prior to the special issue on entheseal changes in the International Journal of Osteoarchaeolology (Henderson and Alves Cardoso, 2013). Nevertheless, their potential should not be disregarded, but further research on normal surface shape variation and size is necessary to understand the relationship between muscle usage, forces at the enthesis and the development of EC. The aim of this paper is to demonstrate the need for quantitative recording methods to achieve these aims, while presenting a pilot study of a method for recording entheses in three-dimensions that is faster than more commonly used methods (Noldner and Edgar, 2013; Nolte and Wilczak, 2013). Problems of reproducibility will also be highlighted.

The interface between two materials which differ in their mechanical characteristics is well-known for being vulnerable, leading to stress concentrating at these zones which can lead to failure (Thomopoulos *et al.*, 2010; Liu *et al.*, 2011). This interface between bone and tendon (or bone and other soft tissue) (Benjamin *et al.*, 2002), i.e. the enthesis, is extremely vulnerable due to the very extreme differences in biomechanical properties (Thomopoulos *et al.*, 2010). Bone is strong in both tension and compression, whereas tendon is only strong in tension (Thomopoulos *et al.*, 2010). Fibrocartilaginous entheses, which are the most well-described, consist of four layers: the soft tissue of the tendon, a layer of unmineralised fibrocartilage, a layer of mineralised fibrocartilage (separated from the layer above by a tidemark, that is not crossed by blood vessels) and the hard tissue that is bone (Benjamin *et al.*, 2002). Their biomechanical characteristics gradually vary, via structural and compositional differences, through the zones of the enthesis (Thomopoulos *et al.*, 2003; Thomopoulos *et al.*, 2010). Entheses often occur on undulating, rounded, or depressed surfaces, therefore forces could become concentrated around specific points, which may lead to points of vulnerability. Modelling of loading has demonstrated that stress is not uniform at idealised versions of these surfaces (Thomopoulos *et al.*, 2006). Describing normal shape variation for specific enthesis is, therefore, vital for the future development of models to characterise these forces and to determine the relationship between EC and stress. To date, there has been limited human osteological research quantifying normal enthesis shape variation (Henderson, 2013).

While there has been limited work solely dedicated to normal variation of enthesis size and shape, there has been a recent increase in the number of studies quantifying, normally in three-dimensions, enthesis size and shape. The study of enthesis area, in either two- or three-dimensions, has been the main focus of research and have included both fibrous and fibrocartilaginous entheses (Wilczak, 1998; Zumwalt, 2006; Pany *et al.*, 2009; Noldner, 2013; Noldner and Edgar, 2013; Nolte and Wilczak, 2013). Other studies have quantified surface shape using methods widely used in materials science, e.g. finite element analysis and shape deviation from planarity (Zumwalt, 2005, 2006; Pany *et al.*, 2009; Henderson, 2013). In general, those studies which analysed the relationship between surface area and EC found an increased area in those with EC (Pany *et al.*, 2009; Noldner, 2013; Noldner and Edgar, 2013). This was also found in two-dimensions (Pany *et al.*, 2009; Henderson, 2013). While one study found that body size correlated with enthesis size (Nolte and Wilczak, 2013) another found no correlation between body size and enthesis size for normal entheses (Henderson, 2013). Surface shape and area analysis demonstrated that muscle size (a direct indicator of strength) is not correlated with enthesis surface complexity (i.e. its height and width variations) in sheep (Zumwalt, 2006). What is not clear from this study is whether any of the sheep studied had EC.

In contrast to the sheep study, no experiments have been undertaken on humans. The closest to these studies are those on identified skeletal collections. These are skeletons of known age, sex and occupation and have been widely used to develop methods for recording entheses (Mariotti *et al.*, 2004; Mariotti *et al.*, 2007; Villotte *et al.*, 2010; Henderson *et al.*, 2013a) and to test them (Alves Cardoso and Henderson, 2010; Milella *et al.*, 2012; Alves Cardoso and Henderson, 2013; Henderson *et al.*, 2013b; Nolte and Wilczak, 2013; Perréard Lopreno *et al.*, 2013). These studies have highlighted the limitations, described above, of the role of ageing in the development of EC, but also the problem of interpreting occupations. Most notably, problems relating to the categorisation of occupations into manual and non-manual work (Alves Cardoso and Henderson, 2013), but also the flexibility of employment and tasks undertaken by men and women throughout their life course (Henderson *et al.*, 2013b). The sheep experiment (Zumwalt, 2006) demonstrated that there was no change in enthesis size or shape in mature individuals, so it is likely that size and shape variation occurs during development. For these reasons, studies on EC should consider the age at which employment began and the tasks, repetition and loading involved.

2. Materials and Methods

The identified skeletal collection in Coimbra encompasses 505 individuals who died between 1904 and 1936, mostly in Coimbra or its environs (Rocha, 1995). While many occupations are represented within the collection (Alves Cardoso and Henderson, 2013), the adult (18+ years old) males (of which there are 266) are most commonly listed as *trabalhador* (Alves Cardoso and Henderson, 2013), of which there are 58 in the collection. The term *trabalhador* refers to labourers or unskilled workers; individuals who mostly worked on the land but also provided services in the cities and towns (Cardoso, 2008). Their socio-economic status would have been low and they are likely to have been involved (particularly those in the countryside) in seasonal work but could also be employed in rural industries, e.g. olive pressing, construction or public works (Fonseca and Guimarães, 2009). In some areas of Portugal, notably Barreiro, this term, during this period, included unqualified work in a large chemical company (Fonseca and Guimarães, 2009). Social mobility was low, with the majority having the same occupation as their father (Fonseca and Guimarães, 2009). It is known from records from the 1960s that the socio-economic group above the *trabalhadores,* skilled metal workers in Lisbon, began work aged between 11 and 14 (67% of these workers) but the majority of these did not start their working lives in the industrial sector (Fonseca and Guimarães, 2009) and it is possible that they began as *trabalhadores*. This occupational category is therefore ideal for studying normal variation in entheses because the occupation listed at death is unlikely to have altered during life and these individuals are likely to have started work before the onset of skeletal maturity, after which we know that size and shape are not altered by activity (Zumwalt, 2006).

Furthermore, this is one of the largest occupational groups represented in the collection.

Two entheses were chosen for this study: the common extensor origin and the biceps brachii insertion. These were selected because they are widely used to study activity, and previous research by the author has focussed on methods to record them (Henderson, 2013; Henderson *et al.*, 2013a). Visual recording of the total sample was undertaken to exclude individuals with skeletal changes indicative of diseases associated with EC, using previously published criteria (Henderson, 2008). While these individuals may have normal entheses, previous studies, albeit limited in sample size, have highlighted that entheses, in these individuals, may differ in size and shape from those in individuals without signs of disease (Henderson, 2013). Visual recording of the common extensor origin and the biceps brachii insertion was used to determine the presence of EC using published criteria (Henderson *et al.*, 2013a) and these entheses were excluded from the sample. Alongside this visual recording, epicondylar breadth and average radial head diameter were also recorded, using digital sliding calipers, to test whether there is a correlation between enthesis area and these body size proxies, as has previously been found for the biceps brachii insertion (Nolte and Wilczak, 2013).

Previous three-dimensional studies of the biceps brachii insertion have studied the entire footprint of the enthesis, including fibrous regions based on dissections (Nolte and Wilczak, 2013). This approach has since been followed without the cadaver dissections (Noldner and Edgar, 2013). While this method captures the totality of the enthesis it does not solely capture the region that is most commonly recorded visually for fibrocartilaginous entheses, i.e. the fibrocartilaginous zone (Villotte, 2010; Henderson *et al.*, 2013a). For this reason only the fibrocartilaginous zones of these two entheses, as have been previously defined (Villotte, 2010), were recorded.

Three-dimensional scanning was undertaken using a NextEngine⁽ᵀᴹ⁾ HD Desktop 3D laser scanner. This scanner uses four solid-state lasers and complementary metal-oxide-semiconductor (CMOS) sensors to record texture when set in the colour mode (Polo and Felicísimo, 2012). This scanner is relatively cheap and portable, but is known to be affected by ambient light conditions, surface colour and reflectivity (Lemes and Zaimovic-Uzunovic, 2009). Following the method described by Nolte and Wilczak (2013), the scanner was set on the macro setting, which gathers the highest resolution data (1016 points per linear cm). Previous studies have described the time-consuming nature of collecting data from multiple angles and stitching these together in the proprietary software, NextEngine™ ® Scan Studio HD ® software (2006–2010) (Noldner and Edgar, 2013). For this reason, this pilot project used only the image from one angle to capture the surface in three-dimensions. This was a test of whether this is a viable method for large-scale accurate data collection and has the further advantage that the scanner and software can be run

Figure 7.1 Image taken with the NextEngine(TM) CMOS camera demonstrating the position of the biceps enthesis on the turntable in relation to the scanner. The image also shows the outline of the fibrocartilaginous portion of the enthesis, (skeleton 2 left biceps brachii).

from a laptop without a dedicated graphics card, thereby reducing costs.

Entheses were outlined using a 0.5 mm HB propelling pencil to ensure an even and clear outline of the enthesis margins and was erased using a soft rubber afterwards. The outline was drawn to ensure that the entire fibrocartilaginous zone was included (Figure 7.1). The radius was placed standing vertically on the turntable and placed so that the biceps brachii enthesis faced the scanner with the interosseous border angled 45 degrees away from the scanner, ensuring that the fibrocartilaginous portion of the enthesis faced the scanner (Figure 7.1). The

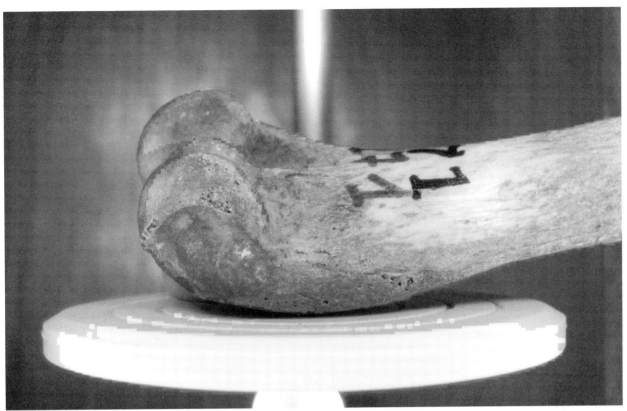

Figure 7.2 Image taken with the NextEngine(TM) CMOS camera demonstrating the position of the common extensor origin on the turntable in relation to the scanner. Image also shows the outline of the fibrocartilaginous portion (skeleton 171 left common extensor origin).

humerus was placed horizontally with the distal condyles perpendicular to the scanner (aligned using a set square) and the distal end was on the turntable with the medial condyle raised so that the lateral supracondylar border faced the scanner (Figure 7.2). Both entheses were placed at 150 mm from the scanner at the closest point. The bones were scanned and the scanned images were cropped along the pencil lines using NextEngine™ ® Scan Studio HD ® software (2006–2010). This software was then used to calculate the area of the enthesis. The point cloud was also exported to allow a two-dimensional measure of surface shape, root mean square (RMS), medio-laterally across the centre of the biceps brachii enthesis and proximo-distally for the common extensor origin following a previously described two-dimensional method (Henderson, 2013). It should be noted that, because this was calculated from the point cloud, this is measured in pixels and therefore the values are not directly comparable to the two-dimensional method. These sections were chosen as they were the fastest to calculate based on the orientation of the bone. RMS calculations were performed in Apache's open source software OpenOffice[TM] 4 version 4.0.1. RMS was only calculated for the left sides of all the skeletons because the pilot nature of this study and limited sample size mean that this is unlikely to contribute greatly to this analysis. Left sides were chosen because the sample size was larger on this side.

One aim of this study was to test the repeatability of this method at each step of the process. As part of this, one female skeleton (skeleton 2) was recorded. Only entheses from the left side were used. Skeleton 2 was used to test variability in the cropping of the data in the software. One scan of each enthesis was taken and cropped five times on the same day, alternating between each enthesis to avoid memorising the resultant shape. The second test was that of alignment of the entheses from skeleton 2. The outlines were kept, but the bones were placed on the turntable ten times on the same day. This involved alternating between scanning each enthesis and taking bone measurements to avoid remembering the exact placement of the bone on the turntable. The final step of this error analysis was redrawing the outlines, rescanning and cropping the bones. This was performed in total five times on the left side of skeleton 2, three times on the same day, once on the following day and the final time seven days later. Outlines were erased immediately after scanning, leaving no trace on the bone, and were only redrawn immediately prior to scanning. It was assumed that the repeatability for RMS and area would be of a similar magnitude, therefore this was only calculated for RMS for the final step.

Recording of the entheses of the *trabalhadores* proceeded over the course of two days following this initial assessment of error rates and using the methods described above. To test the repeatability of the method over time, three skeletons were re-recorded over eighty days later. These were the left and right sides of skeletons 2 and 118, the latter being chosen at random from the *trabalhadores*.

Repeatability was calculated using percent error (Nolte and Wilczak, 2013) calculated as follows:

$$percent\ error = \frac{n_{max} - n_{min}}{\bar{n}}$$

in OpenOffice version 4.0.1. The R Package version 2.15.1 was used to calculate all other statistical tests and plot data. Where outliers (outside the 1.5 interquartile range) were found, percent errors were recalculated without the outlier. For the remaining data set, all data were tested for normality using a Shapiro-Wilk test. For comparisons between linear measurements and area, the area data were $\log_{10}$ transformed prior to analysis (Nolte and Wilczak, 2013) and prior to tests of normality being undertaken. One-tailed Spearman's rank correlations were used to test whether the area of entheses were symmetrical and whether they correlated with body size proxies. The hypothesis was that these would increase proportionally. A two-tailed Spearman test was used to determine whether age was correlated with area, to avoid assuming directionality. It should be noted that log transformation was not necessary once it was determined that the data were non-normally distributed, therefore rank order tests had to be performed. However, the log transformations are presented to enable comparisons with other studies. Partial correlations were not performed, due to the limited sample size. Statistical significance was set at 95%. Scatterplots of the results with marginal boxplots were used to interpret these results.

3. Results

Of the 58 *trabalhadores* in the collection only 24 individuals met the sampling requirements. These individuals ranged in age from 18 to 60, with the majority of individuals under the age of 40 and with a median age of 36 years. This sample is small and was reduced further for individual entheses and measurements due to damage or the presence of EC. Sample sizes for each variable can be seen in Table 7.1.

3.1. Repeatability

Table 7.2 demonstrates that, even in the initial phase of cropping a scan, error creeps in, with a range of error from 2.9% in the biceps brachii insertion to 6.1% in the common extensor origin. However, it should be noted that one scan for the common extensor origin was accidentally rotated in the software and appears to have been misaligned. This is an outlier. Without this outlier the error for this enthesis is 3.4%. For the tests of alignment the error is lower in the common extensor origin than the biceps brachii insertion (Table 7.2), but both have outliers. It should be remembered that the cropping error also affects this measure. The final test of repeatability of each step in the scanning process was redrawing the outlines of each enthesis. This produced a high error rate (Table 7.2), which was reduced by removing the data from seven days after the initial test. For the common extensor origin, this value was an outlier, but it was not for the biceps brachii. In doing this, the error was reduced for the common extensor

Table 7.1 Descriptive statistics and correlations between age, area and body size. Abbreviations: b.=brachii, ext.=extensor.

Variable 1	n	min	max	median	σ	Variable 2	n	min	max	median	σ	Spearman's rank correlations between variables 1 and 2 — n	rho	p
Age	24	18	60	36	13	Log10 transformed left biceps b. area	see values for variable 1					15	-0.15	0.60
						Log10 transformed right biceps b. area						13	0.30	0.31
						Log10 transformed left common ext. area						11	-0.46	0.15
						Log10 transformed right common ext. area						11	0.46	0.16
Left biceps b. area	15	60.27	133.51	94.97	22.01	Right biceps b. area	13	57.88	190.55	101.32	43.38	7	0.25	0.30
Left common ext. area	11	90.77	152.68	124.27	17.06	Right common ext. area	11	85.09	185.11	135.2	29.11	7	-0.21	0.70
Log10 transformed left biceps b. area	15	1.78	2.13	1.98	0.10	Left humeral epicondyle width	17	54.84	67.95	61.89	3.45	13	0.04	0.45
Log10 transformed right biceps b. area	13	1.76	2.28	2.01	0.17	Right humeral epicondyle width	17	55.76	67.18	61.54	3.11	11	0.59	0.03
Log10 transformed left common ext. area	11	1.96	2.18	2.09	0.06	Left average radial head diameter	10	19.05	26.85	22.71	1.94	7	0.07	0.45
Log10 transformed right common ext. area	11	1.93	2.27	2.13	0.10	Right average radial head diameter	11	19.15	24.36	22.64	1.41	10	-0.30	0.81
Log10 transformed left biceps b. area	see values above					Left biceps b. RMS	15	24.94	34.06	32.27	2.99	15	0.03	0.46
Log10 transformed left common ext. area						Left common ext. RMS	11	14.95	35.23	29.73	6.06	11	-0.02	0.53

Table 7.2 Repeatability of the scanning steps for area calculation (areas in mm²): descriptive statistics and percent error.

Steps	Common extensor origin					Biceps brachii				
	n	Min	Max	Mean	% error	n	Min	Max	Mean	% error
Cropping	5	67.17	71.43	69.45	6.12	5	65.54	67.45	66.52	2.87
Alignment	10	67.34	70.01	69.23	3.87	10	64.80	68.61	66.26	5.75
Outline redrawing	5	68.79	94.14	74.37	34.08	5	63.18	85.24	69.94	31.53

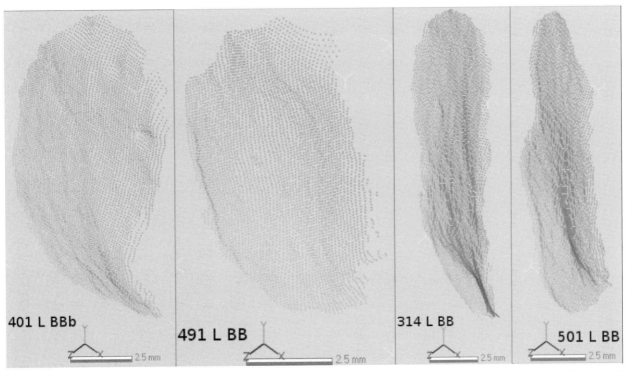

Figure 7.3 Variation in enthesis shape demonstrated by the point clouds of data, representing entheses in the same plane, in Geomagic® Verify(TM) Viewer 2014 version 4.0.0.0. The variety of shapes demonstrates the lowest RMS values (skeleton 401 aged 24, RMS=24.94 area=96.04mm²) to the highest (skeleton 501 aged 27, RMS=34.06 area=85.11 mm²) alongside the second lowest (skeleton 491 aged 18, RMS=25.11, area=70.51 mm²) and second highest (skeleton 314 aged 34, RMS=33.97 area=112.88 mm²)

origin to 1.8% and to 5.2% for the biceps brachii. For the RMS values (Table 7.3) only the final step was tested and the error was found to be 15.3% and 26.3% for the biceps brachii and common extensor origin respectively; no outliers were present. Bone measurements (epicondylar width and average radial head diameter) were also tested for repeatability with 0.2% and 2.1% error, respectively.

The test of repeatability through time demonstrated high levels of variability in error and the visual recording of the enthesis (Table 7.4). One enthesis was recorded the first time, but during the second observation it was

Table 7.3 Repeatability of scanning and its effect on RMS values: descriptive statistics and percent error for RMS scores. Data collected from the same data set, and therefore comparable to that used to calculate the percent error in the bottom line of Table 7.2.

Enthesis	n	Min	Max	Mean	% error
Common extensor origin	5	27.58	35.65	30.71	26.27
Biceps brachii	5	29.49	34.36	31.77	15.33

determined to be damaged and not suitable for recording. Aside from this difference, the error ranged from 21% to 40% for the common extensor origin and 7% to 65% for the biceps brachii insertion.

3.2. Size and Shape Variability

Descriptive statistics (Table 7.1) demonstrate that there is wide variation in enthesis area for both the common extensor origin and biceps brachii entheses, with a minimum range of 61.9 mm² in the case of the left common extensor origin and a maximum range of 132.7 mm² in the case of the right biceps brachii. The standard deviations also demonstrate that the right side has a greater variability than the left side and that this variability is higher in the biceps brachii than the common extensor origin. Shape variability, measured by RMS, was only calculated for the left side of both entheses. In contrast to the area, there is greater range and variability in the common extensor origin RMS. Figure 7.3 demonstrates the minimum and maximum RMS values for the biceps brachii insertion and also highlights the variation in area.

Table 7.4 Repeatability through time for area calculation (areas in mm²): descriptive statistics and percent error.

Skeleton	Common extensor origin			Biceps brachii insertion		
	1st observation (area in mm²)	2nd observation (area in mm²)	% error	1st observation (area in mm²)	2nd observation (area in mm²)	% error
2 left	70.01	88.14	22.93	64.80	104.68	47.06
2 right	67.66	damaged	-	46.27	91.29	65.45
118 left	110.65	137.01	21.28	112.16	132.05	16.29
118 right	98.43	147.11	39.65	159.47	148.16	7.35

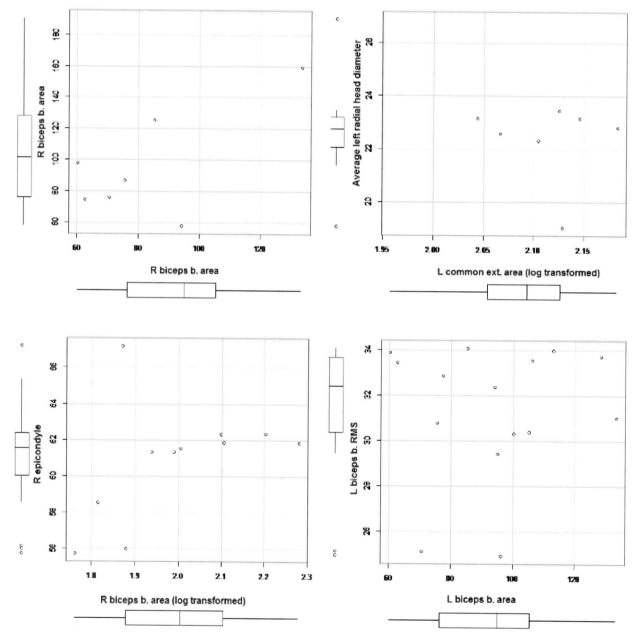

Figure 7.4 Scatterplots with marginal boxplots pinpointing the outliers of four of the associations tested.

Tests for correlations between the areas of left and right sides of entheses demonstrated no significant correlations (Table 7.1), nor were there any correlations with age-at-death. The only correlation found was between the $\log_{10}$ transformed right biceps brachii area and the distal epicondyle of the same side (p=0.03). Figure 7.4 shows some of the scatterplots with marginal boxplots indicating outliers from the 1.5 interquartile range. As can be seen from these plots, the body proxies all have outliers and there are no clear associations between the variables. This is also the case for the RMS values, which do not correlate with the areas of the entheses studied (Table 7.1).

4. Discussion

Three-dimensional methods for recording entheses are becoming more common, but one of their limitations is their time-consuming nature, both at the initial scanning stage and subsequent cropping of the scans for data analysis. One aim of this pilot study was to test whether enthesis area could be captured from one angle, thereby reducing the scanning time by at least a third compared to the time required for a three-angled or longer for a 360° scan. The limitation of a single-angled scan is that the enthesis must be completely within the viewing angle of the scanner and, to ensure repeatability and comparability; this angle must be the same each time. Repeatability for the multi-angled approaches ranges from 10% to 16.5% (Noldner, 2013; Nolte and Wilczak, 2013) and it is clear from the data presented in this paper that the single-angled approach has a lower rate of repeatability with percent errors ranging from 7% to 65.5%. However, this error was measured over a period of over 80 days compared to the 'several days' of one study and approximate two weeks of another (Noldner, 2013; Nolte and Wilczak, 2013). Other factors contributing to this, aside from the methodology, include differences in lighting, which is known to affect this scanner (Lemes and Zaimovic-Uzunovic, 2009). A further problem was the definition of the fibrocartilaginous footprints. In the 80-day period between the first and final error analysis, an online discussion was organised to test inter-observer error of the 'Coimbra method' (Henderson *et al.*, 2013a). This led to a discussion about the footprints of fibrocartilaginous entheses and, despite being based on the definition used here (Villotte, 2008), will have influenced the outlining of entheses on the second recording.

It is clear that the limitations described above are not the sole problems. Error creeps in at each step of the process. Although cropping the image is not the first recording point, this is the only step not influenced by other factors. One scan was taken of each enthesis and this was repeatedly cropped alternating between the two entheses to reduce the effect of memorising the shape achieved after the cropping. During this step there was limited but still important error ranging from 2.9% to 6.1%. The step itself involves cropping the image, using the proprietary software, along the pencil drawn lines. It is possible that chalk used in other studies (Noldner, 2013; Nolte and Wilczak, 2013) is more reflective, which may have made this step easier and reduced error. However, during testing of the method, prior to this analysis, it was clear that following a very well-defined and accurate line was important to reduce error. The fine lead pencil, used in this study, was used to achieve this.

Aligning the bones was one of the biggest challenges. The location of an enthesis on a bone can differ between individuals, so a method had to be defined to minimise the effect of individual variation on what the scanner would capture of an enthesis during scanning. It was determined that replicating previous three-dimensional scans of the biceps brachii enthesis (Noldner, 2013; Nolte and Wilczak, 2013) using a single-angled scan would be impossible, but

this was not the goal of this study. The goal was to focus on the fibrocartilaginous footprint, which is the area recorded in most visual methods (e.g. Villotte, 2006; Henderson *et al.*, 2013a). All alignment tests were performed on one day, which is likely to have reduced error rates. However, alignment of the common extensor origin to enable the entire enthesis to be recorded in a single-angled scan was impossible in many cases, particularly in those in which the footprint extends proximally from the lateral epicondyle. This highlights an important limitation of the method, which should be considered when interpreting the data from the main sample. It must be remembered that this problem only affects the common extensor origin and neither of the three-dimensional comparison studies recorded this enthesis (Noldner and Edgar, 2013; Nolte and Wilczak, 2013). Despite this, the results of the correlations produce similar results for both entheses in this study.

The test of the cumulative impact of error from each step of the recording process was that involving the redrawing of the outlines of each enthesis. This step had high error rates (over 30%) for both entheses. This final step demonstrated the error from redrawing the outlines on the bones, but also incorporated the alignment of the bones on the turntable and the effect of cropping each scan. It is therefore not surprising that this step has the highest error. The time scale for this step was also the largest and this is also likely to have increased error rates. This set of scans were also used to calculate error for the RMS scores. For the biceps brachii this error was less than half that of the area, while that for the common extensor origin was 26%. This demonstrates, as is unsurprising, that area calculations are more susceptible to minimal changes in scanning than two-dimensional calculations from the same dataset.

While the errors from the scans are large, they should be placed in perspective in relation to other methods for recording entheses. A recent method tested by its creators had a percentage agreement of only 71.8%, but for some entheses and features was as low as 52.5% (Henderson *et al.*, 2013a). An independent test of the, until recently, most widely used method of recording entheses demonstrated percentage agreements of between 32% and 92% (Davis *et al.*, 2013). For the biceps, the agreement ranged from 48% to 58% depending on trait scored, while for the common extensor origin the agreement ranged from 49% to 65% (Davis *et al.*, 2013). Although these studies all report inter-observer agreement as opposed to intra-observer error, these results demonstrate the need for clearer definitions of recording methods and, most importantly in relation to three-dimensional studies, clear definitions of the footprints of entheses on bones. Most importantly, it highlights the need to work together on developing methods that can be widely applied, because current error rates indicate that individual studies are not comparable (Davis *et al.*, 2013).

Error is clearly a problem for this study and should not be ignored, but some of this has crept in from the time-frame for the error analysis, which is considerably longer than in other studies. Recording error will have affected the recording of the main sample of skeletons, but it should be borne in mind that these were recorded successively over the course of a very short time-frame, which may have reduced this problem. The aim of this section of the pilot project was to study normal enthesis size and shape variation in a group of individuals who would have started their working lives before skeletal maturity and whose occupation is likely to have been static throughout life. However, the only group of individuals in the Coimbra skeletal collection who satisfied these requirements were the *trabalhadores,* who were unskilled labourers. It is important to remember that they are likely to have performed many and varied tasks and that these may not have overlapped between individuals. Therefore they do not represent a cohesive group and this, as well as error, is a significant limitation of this study.

The *trabalhadores* with normal entheses were mostly young, under the age of forty. This is unsurprising as the frequency of entheseal changes is known to increase with age (Milella *et al.*, 2012; Alves Cardoso and Henderson, 2013). Less than half of the original sample group met the criteria for analysis or had damaged entheses, therefore the sample size for this study was smaller than originally anticipated. The research question related not only to the variation in area and shape, but also to their relationship with body size and age. Variation in area was greatest in the right side for both entheses. Previous studies have demonstrated no statistically significant differences between mean left and right entheseal areas in the biceps brachii (Nolte and Wilczak, 2013), but the sample size in this study made it inappropriate to test this. However, no correlations between left and right sides were found; this could be a result of sample size. A previous study did find a correlation between right and left biceps brachii in a similar sized sample (Noldner and Edgar, 2013). However, neither of these studies focussed solely on the fibrocartilaginous area. This pilot study demonstrates side asymmetry in the fibrocartilaginous portion, which may indicate that Noldner and Edgar's results are caused by symmetry in the fibrous portion of this enthesis. Differences in the results of the correlations between this study and others (Noldner and Edgar, 2013; Nolte and Wilczak, 2013), described below, could also be a result of this effect and should be tested further. Whether these results are an effect of biological differences or differences in childhood work patterns, which may have made this sample more asymmetric, could not be determined but should be tested in the future.

The shape of entheses was also measured on the left sides and, had it been known in advance that area was considerably more variable on the right side, then RMS would have been calculated on both sides. However, this was a pilot study aimed at highlighting differences. What is clear is that there is variability and, more importantly that this does not correlate with enthesis area. This has not previously been tested, but is unsurprising. RMS measures the amplitude of enthesis shape by calculating its standard deviation from a mean line, thus a lower value indicates a flatter surface (Henderson, 2013). Area, which can be affected by the amplitude of the enthesis shape, is also affected by other factors, e.g. outline shape and size, which do not affect the RMS calculation. Implementing this measure, or other measures of surface shape, e.g. finite element analysis (Zumwalt, 2005), should be undertaken more frequently alongside the measure of area.

Enthesis area has previously been found to correlate with body size proxies and age (Nolte and Wilczak, 2013). In this study only one correlation was found, between the right humeral epicondyle and the right biceps brachii area. Sample size is likely to have played a role in this as can be seen from the graphs (Figure 7.4). The difference between this study and that described above is likely to be an effect of sample size. However, other factors also contribute to these findings. Unlike previous three-dimensional studies, this study focussed solely on individuals without entheseal changes. Entheseal changes, particularly those which involve bone formation at the margins of entheses, i.e. zone one of the 'Coimbra method' (Henderson *et al.*, 2013a), increase enthesis surface area. This has been demonstrated in previous quantitative studies (Pany *et al.*, 2009; Henderson, 2013; Nolte and Wilczak, 2013). The correlation between enthesis area and age, found in one study, was ascribed to the presence of new bone formation in the older individuals (Nolte and Wilczak, 2013).

It is clear from the intra-observer error that this method is inappropriate for the recording of entheses in three-dimensions. However, what is clear from studies of inter-observer repeatability for visual recording methods is that developing a recording method that can be widely applied by other researchers has not yet been achieved. It is also likely that the three-dimensional studies, which have been published, are subject to similarly high levels of inter-observer error, but such analyses have not been reported. While the error rates are high and inferences based on the remaining data set should therefore be considered extremely cautiously, this study has found that normal fibrocartilaginous footprints of two entheses are variable in both size and shape and that there seems to be no correlation between left and right sides. This variability within (left and right sides) and between individuals could indicate that loading patterns during development may determine the size and shape of the enthesis; it is also possible that this may lead some individuals to be more vulnerable to stress-related changes. It is clear that further testing on a larger sample size with a more robust recording methodology is required to interpret these results and test these hypotheses.

5. Conclusions

This pilot project aimed to test a fast three-dimensional recording method for the fibrocartilaginous footprint of two entheses and to determine variation in size and shape of normal entheses in a well-defined occupational category who would have begun manual work before skeletal maturity. The error rates for the method are high, but this has been found to be a problem for other recording methods. Nevertheless these rates are higher than normally found for intra-observer error analyses. It is therefore clear that the method itself, while fast, is not robust. Although the total sample of *trabalhadores*, unskilled labourers, used in this study was large, the selection criteria for this study, particularly the need for normal entheses, reduced the sample size by more than half. The final distribution supports other research which has shown that normal entheses are mostly found in those under forty years of age. The results of this study contrast with those of other three-dimensional studies of the biceps brachii which demonstrate correlations in area between left and right sides; enthesis area and body size proxies; and enthesis area and age (Noldner, 2013; Noldner and Edgar, 2013; Nolte and Wilczak, 2013). There are several possible reasons for these results which include: small sample size in this study; weak recording methodology; the difference in outlines used; but perhaps most importantly the focus in this study on one occupational group who began work before skeletal maturity and whose entheses were normal. Further research is necessary, using a more robust method, to test this.

Acknowledgements

The author would like to acknowledge the postdoctoral fellowship from Fundação para a Ciência e a Tecnologia (reference: SFRH/BPD/82559/2011) and funding from CIAS – Research Centre for Anthropology and Health, Coimbra to enable attendance at the 15th Annual Meeting of the British Association for Biological Anthropology and Osteoarchaeology in York.

References

Alves Cardoso, F., Henderson, C.Y. 2010. Enthesopathy formation in the humerus: Data from known age-at-death and known occupation skeletal collections. *American Journal of Physical Anthropology* 141, 550–560.

Alves Cardoso, F., Henderson, C. 2013. The categorisation of occupation in identified skeletal collections: A source of bias? *International Journal of Osteoarchaeology* 23, 186–196.

Benjamin, M., Kumai, T., Milz, S., Boszczyk, B.M., Boszczyk, A.A., Ralphs, J.R. 2002. The skeletal attachment of tendons - tendon 'entheses'. *Comparative Biochemistry and Physiology Part A* 133, 931–945.

Cardoso, F.A. 2008. A portrait of gender in two 19th and 20th century Portuguese populations: A palaeopathological perspective. Unpublished PhD thesis, Department of Archaeology, Durham University, Durham.

Davis, C.B., Shuler, K.A., Danforth, M.E., Herndon, K.E. 2013. Patterns of interobserver error in the scoring of entheseal changes. *International Journal of Osteoarchaeology* 23, 147–151.

Fonseca, H.A., Guimarães, P.E. 2009. Social mobility in Portugal (1860–1960): Operative issues and trends. *Continuity and Change* 24, 513.

Foster, A., Buckley, H., Tayles, N. 2012. Using enthesis robusticity to infer activity in the past: A review. *Journal of Archaeological Method and Theory*, 1–23.

Godde, K., Taylor, R. W. 2011. Musculoskeletal stress marker (MSM) differences in the modern American upper limb and pectoral girdle in relation to activity level and body mass index (BMI). *Forensic Science International* 210, 237–242.

Hawkey, D.E., Merbs, C.F. 1995. Activity-induced musculoskeletal stress markers (MSM) and subsistence strategy changes among ancient Hudson Bay eskimos. *International Journal of Osteoarchaeology* 5, 324–338.

Henderson, C.Y. 2008. When hard work is disease: the interpretation of enthesopathies, in: Brickley, M., Smith, M. (Eds.), *Proceedings of the Eighth Annual Conference of the British Association for Biological Anthropology and Osteoarchaeology*. British Archaeological Reports, International Series 1743, Oxford, pp. 17–25.

Henderson, C.Y. 2013. Technical note: Quantifying size and shape of enthuses. *Anthropological Science* 121, 63–73.

Henderson, C.Y., Alves Cardoso, F. 2013. Preface to special issue entheseal changes and occupation: Technical and theoretical advances and their applications. *International Journal of Osteoarchaeology* 23, 127–134.

Henderson, C.Y., Caffell, A.C., Crapps, D.D., Millard, A.R., Gowland, R. 2013b. Occupational mobility in nineteenth century rural England: The interpretation of entheseal changes. *International Journal of Osteoarchaeology* 23, 197–210.

Henderson, C.Y., Mariotti, V., Pany-Kucera, D., Villotte, S., Wilczak, C.A. 2013a. Recording specific features of fibrocartilaginous entheses: Preliminary results of the Coimbra standard method. *International Journal of Osteoarchaeology* 23, 152–162.

Humphries, J., Sarasúa, C. 2012. Off the record: Reconstructing women's labor force participation in the European past. *Feminist Economy* 18, 39–67.

Jurmain, R.D., Alves Cardoso, F., Henderson, C.Y., Villotte, S. 2012. Bioarchaeology's Holy Grail: The reconstruction of activity, in: Grauer, A. (Ed.), *Companion to Paleopathology*. Wiley-Blackwell, Chichester, pp. 531–552.

Lemes, S., Zaimovic-Uzunovic, N. 2009. Study of ambient light influence on laser 3D scanning. 7th int. conference on industrial tools and material processing technologies. Ljubljana, Slovenia.

Lieverse, A.R., Stock, J. T., Katzenberg, M. A., and Haverkort, C. M. 2011. The bioarchaeology of habitual activity and dietary change in the Siberian Middle Holocene, in: Pinhasi, R., Stock, J.T. (Eds.), *Human Bioarchaeology of the Transition to Agriculture*. John Wiley and Sons, Ltd., Chichester, pp. 263–291.

Liu, Y., Birman, V., Chen, C., Thomopoulos, S., Genin, G.M. 2011. Mechanisms of bimaterial attachment at the interface of tendon to bone. *Journal of Engineering Materials and Technology* 133.

Mariotti, V., Belcastro, M.G. 2011. Lower limb entheseal morphology in the Neandertal Krapina population (Croatia, 130 000 BP). *Journal of Human Evolution* 60, 694–702.

Mariotti, V., Facchini, F., Belcastro, M.G. 2004. Enthesopathies – proposal of a standardized scoring method and applications. *Collegium Anthropologicum* 28, 145–159.

Mariotti, V., Facchini, F., Belcastro, M.G. 2007. The study of entheses: Proposal of a standardised scoring method for twenty-three entheses of the postcranial skeleton. *Collegium Anthropologicum* 31, 291–313.

Milella, M., Belcastro, M.G., Zollikofer, C.P., Mariotti, V. 2012. The effect of age, sex, and physical activity on entheseal morphology in a contemporary Italian skeletal collection. *American Journal of Physical Anthropology* 148, 379–388.

Myszka, A., Piontek, J. 2011. Shape and size of the body vs. musculoskeletal stress markers. *Anthropologischer Anzeiger* 68, 139–152.

Myszka, A., Piontek, J. 2012. Variation of musculoskeletal stress markers in the medieval population from Cedynia (Poland) – proposal of standardized scoring method application. *Collegium Anthropologicum* 36, 1009–1017.

Niinimäki, S. 2012. The relationship between musculoskeletal stress markers and biomechanical properties of the humeral diaphysis. *American Journal of Physical Anthropology* 147, 618–628.

Noldner, L. 2013. *Spanish Missionization and Maya Social Structure: Skeletal Evidence for Labor Distribution at Tipu, Belize.* University of New Mexico, Albuquerque.

Noldner, L.K., Edgar, H.J.H. 2013. 3D representation and analysis of enthesis morphology. *American Journal of Physical Anthropology* 152, 417–424.

Nolte, M., Wilczak, C. 2013. Three-dimensional surface area of the distal biceps enthesis, relationship to body size, sex, age and secular changes in a 20th century American sample. *International Journal of Osteoarchaeology* 23, 163–174.

Pany, D., Viola, T., Teschler-Nicola, M. 2009. The scientific value of using a 3D surface scanner to quantify entheses, workshop in musculoskeletal stress markers (MSM): Limitations and achievements in the reconstruction of past activity patterns, July 2–3, 2009. Coimbra, CIAS – Centro de Investigação em Antropologia e Saúde. http://www.uc.pt/en/cia/msm/MSM_podium (September, 2013).

Perréard Lopreno, G., Alves Cardoso, F., Assis, S., Milella, M., Speith, N., 2013. Categorization of occupation in documented skeletal collections: Its relevance for the interpretation of activity-related osseous changes. *International Journal of Osteoarchaeology* 23, 175–185.

Peterson, J., Hawkey, D.E. 1998. Preface. *International Journal of Osteoarchaeology* 8, 303–304.

Polo, M.-E., Felicísimo, Á.M. 2012. Analysis of uncertainty and repeatability of a low-cost 3D laser scanner. *Sensors* 12, 9046–9054.

Rocha, M.A. 1995. Les collections ostéologiques humaines identifiées du Musée Anthropologique de l'Université de Coimbra. *Antropologia Portuguesa* 13, 7–38.

Rojas-Sepúlveda, C.M., Rivera-Sandoval, J., and Martín-Rincón, J. G. 2011. Paleoepidemiology of pre-Columbian and Colonial Panamá Viejo: A preliminary study. *Bulletin et Memoires de la Societe d'Anthropologie de Paris* 23, 70–82.

Santos, A.L., Alves-Cardoso, F., Assis, S., Villotte, S. 2011. The Coimbra workshop in musculoskeletal stress markers (MSM): An annotated review. *Antropologia Portuguesa* 28, 135–161.

Schlecht, S.H. 2012a. A histomorphometric analysis of muscular insertion regions: Understanding enthesis etiology. Department of Anthropology, Ohio State University.

Schlecht, S.H. 2012b. Understanding entheses: Bridging the gap between clinical and anthropological perspectives. *Anatomical Record* 295, 1239–1251.

Shuler, K.A., Hodge, S.C., Danforth, M.E., Lynn Funkhouser, J., Stantis, C., Cook, D.N., Zeng, P. 2012. In the shadow of Moundville: A bioarchaeological view of the transition to agriculture in the central Tombigbee valley of Alabama and Mississippi. *Journal of Anthropological Archaeology* 31, 586–603.

Stefanović, S., Porčić, M. 2011. Between-group differences in the patterning of musculo-skeletal stress markers: Avoiding confounding factors by focusing on qualitative aspects of physical activity. *International Journal of Osteoarchaeology* 23, 94–105. DOI: 10.1002/oa.1243.

Thomopoulos, S., Genin, G.M., Galatz, L.M. 2010. The development and morphogenesis of the tendon-to-bone insertion – what development can teach us about healing. *Journal of Musculoskeletal and Neuronal Interactions* 10, 35–45.

Thomopoulos, S., Marquez, J.P., Weinberger, B., Birman, V., Genin, G.M. 2006. Collagen fiber orientation at the tendon to bone insertion and its influence on stress concentrations. *Journal of Biomechanics* 39, 1842–1851.

Thomopoulos, S., Williams, G.R., Gimbel, J.A., Favata, M., Soslowsky, L.J. 2003. Variation of biomechanical, structural, and compositional properties along the tendon to bone insertion site. *Journal of Orthopaedic Research* 21, 413–419.

Üstündağ, H., Deveci, A. 2011. A possible case of Scheuermann's disease from Akarçay Höyük, Birecik (Şanlıurfa, Turkey). *International Journal of Osteoarchaeology* 21, 187–196.

Villotte, S. 2006. Connaissances Médicales Actuelles, Cotation des Enthésopathies: Nouvelle Méthode. *Bulletin et Memoires de la Societe d'Anthropologie de Paris* 18, 65–85.

Villotte, S. 2008. Enthésopathies et activités des hommes préhistoriques-Recherche méthodologique et application aux fossiles européens du Paléolithique supérieur et du Mésolithique. Université Sciences et Technologies-Bordeaux I, Bordeaux.

Villotte, S. 2010. Actualité de la recherche: les altérations de l'enthèse, pourquoi et comment les reconnaître. Actualité de la recherche: les altérations de l'enthèse, pourquoi et comment les reconnaître. Workshop de la Société Suisse d'Anthropology.

Villotte, S., Castex, D., Couallier, V., Dutour, O., Knüsel, C.J., Henry-Gambier, D. 2010. Enthesopathies as occupational stress markers: evidence from the upper limb. *American Journal of Physical Anthropology* 142, 224–234.

Villotte, S., Knüsel, C.J. 2012. Comment on "Musculoskeletal stress marker (MSM) differences in the modern American upper limb and pectoral girdle in relation to activity level and body mass index (BMI)" by K. Godde and R. Wilson Taylor. *Forensic Science International* 217, e31.

Villotte, S., Knüsel, C.J. 2013. Understanding entheseal changes: Definition and life course changes. *International Journal of Osteoarchaeology* 23, 135–146.

Wilczak, C.A. 1998. Consideration of sexual dimorphism, age, and asymmetry in quantitative measurements of muscle insertion sites. *International Journal of Osteoarchaeology* 8, 311–325.

Zumwalt, A.C. 2005. A new method for quantifying the complexity of muscle attachment sites. *Anatomical Record (Part B) 286B*, 21–28.

Zumwalt, A.C. 2006. The effect of endurance exercise on the morphology of muscle attachment sites. *The Journal of Experimental Biology* 209, 444–454.

A Biocultural Assessment of Health in Roman York

Lauren McIntyre and Andrew T. Chamberlain

This study combines new and pre-existing osteological evidence with archaeological evidence in order to assess the health status of the population of Roman York. This research also examines differences in health according to social and occupational status categories within the population and compared to other contemporary Romano-British urban populations. Results indicate that York had significantly elevated prevalence of ante-mortem trauma, and os acromiale. Prevalence of os acromiale, Schmorl's nodes, non-specific infection and trauma were also observed as being significantly different between males and females from York. Distribution of os acromiale was also found to be significantly different between the left and right scapulae of individuals from York. Elevated prevalence of traumatic injury in some skeletal elements of the cranium and several post-cranial skeletal elements was significantly associated with an unusual group of burials from sites located on Driffield Terrace.

Keywords Roman Britain; Palaeopathology; Trauma; Infection; Driffield Terrace

1. Introduction

Britain's first direct contact with the Roman Empire came with Julius Caesar's invasions in 55 and 54 BC, after which a certain amount of trade and diplomatic links were established (Mattingly, 2007: 64–67). Colonisation of Britain and development of 'Romanised' urban centres, however, did not begin until after the large scale military invasion of AD 43, under the Emperor Claudius (De la Bédoyère, 2003: 15–8). The end of the Roman occupation of Britain is commonly considered to be around AD 409-410, when the Britons revolted against Roman rule (Mattingly, 2007: 529–531).

The establishment (in the 1st century AD) of a Roman legionary fortress at *Eboracum* (present-day York) was followed by the development of a civilian settlement, sometimes referred to as the *canabae*. In the early 3rd century AD the town was given official *colonia* status, which also gave the inhabitants of the town Roman citizenship (Ottaway, 2004). Both the military and civilian sectors of the population of York were served by a variety of large and small cemeteries distributed around the approach roads principally to the north, south-east and south-west of the fortress (RCHME, 1962: 67; Jones, 1984: 34). Burials in all cemeteries are represented by inhumation and cremation burials, spanning the full period of Roman occupation from the 1st–4th centuries AD and including the remains of both military and civilian personnel.

1.1. Excavation of the Cemeteries of Eboracum

The two best-known excavations of the cemeteries of Roman York took place at the Old Railway Station and at Trentholme Drive (Figure 8.1). Large concentrations of burials dating to the second to fourth centuries were discovered during the construction of York Railway Station and its associated works (e.g. goods lines, engine sheds etc.) in the 19th century (RCHME, 1962: 77). Romano-British human skeletal remains were excavated at Trentholme Drive by Leslie Peter Wenham and the Ministry of Works between 1951 and 1959, where a minimum of 350 individuals from inhumation burials were excavated, as well as a further 53 cremation burials (Wenham, 1968).

Another excavated cemetery area, situated on the main road leaving York to the southwest, is referred to as The Mount. This area comprises several archaeological sites,

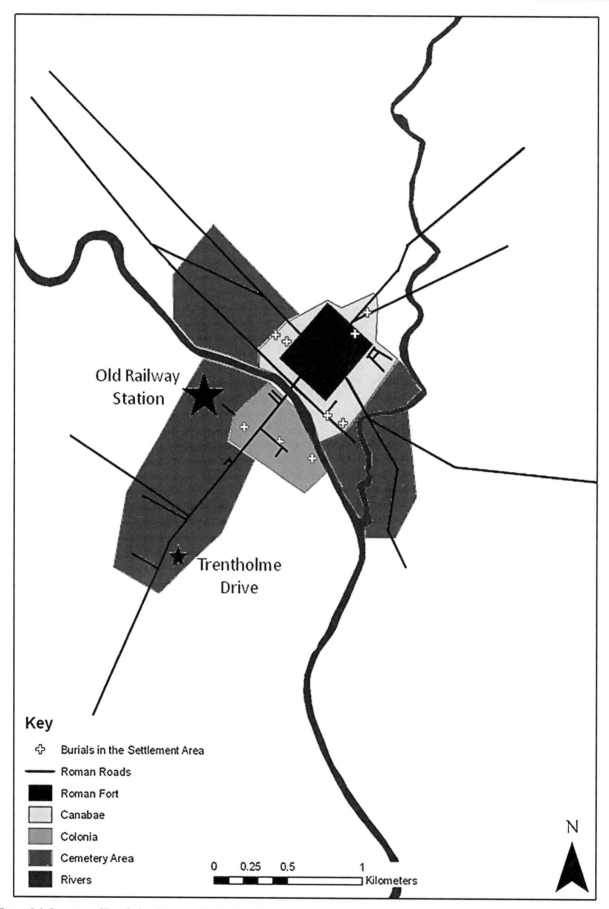

Figure 8.1 Location of Trentholme Drive and the Railway Station cemeteries, situated around the fortress at Eboracum (McIntyre, 2013).

including Micklegate, Blossom Street, Driffield Terrace/ Estate, Trentholme Drive, Holgate and Dringhouses, and has been excavated in a rather piecemeal fashion since the 19th century (RCHME, 1962: 92-107). The Driffield Terrace region of The Mount cemetery has recently been subject to a significant amount of research, due to the discovery of a large group of decapitated burials (Tucker, 2006; Montgomery *et al.*, 2011: 141, 148; York Archaeological Trust, 2011).

As well as burials found in the main cemetery/ excavated areas, single or small clusters of Roman burials are periodically found in other areas of York. These are often found on small-scale excavations, evaluations or watching briefs, but nevertheless add to the growing corpus of osteological data relating to the former inhabitants of *Eboracum*.

1.2. Past Research on Health in Eboracum

Studies seeking to reconstruct health patterns across the Roman population have been limited to the assessment of health in the Trentholme Drive skeletal sample by Warwick (1968), a reconstruction of public health in York from the Roman period until the 18th century by Addyman (1989), and more recently, Peck's (2009) doctoral study examining the biological and sociocultural effects of urbanisation associated with Roman colonialism on the overall health of the population. This limited body of research presents a stark contrast to the number of studies undertaken on the health of the medieval population of York (e.g. Grauer, 1993; Lewis *et al.*, 1995; Grauer and Roberts, 1996; Lewis, 2002; Sullivan, 2005; Watts, 2011).

Warwick's (1968: 158–162) study of the Trentholme Drive assemblage found no evidence of specific infectious, malignant neoplastic or metabolic disease, but a high incidence osteoarthritis and ante-mortem fracture (apparently only found in males), leading to the conclusion that the population were extremely physically hard working and that the male component of the population were most likely to have been soldiers. Addyman's (1989: 252) review concluded that the health of the population was relatively good, though the osteological findings largely reference Warwick's results from Trentholme Drive. Peck's (2009) results showed that the later Romano-British skeletal sample had significantly higher rates of pathological conditions indicative of non-specific physiological stress than their pre-colonial counterparts, indicating that Roman colonialism had a substantial impact on the health status of the population. While several unpublished reports have examined and discussed observed pathological lesions on individual or small groups of human skeletons (e.g. McIntyre and Holst, 2006; Sulosky, 2006; McIntyre, 2007; Holst, 2008, 2009, 2010, 2011) there has been no attempt to collate these findings until the present project.

1.3. Health in Roman Britain

Roberts and Cox's (2003) review of the history of health in Britain surmised that the development of the first truly urban centres (as well as other factors such as diversification in industry, agriculture and trade, increase in the range of available foodstuffs etc.) during the Roman period resulted in changes in health across the British population. The prevalences of observed pathological conditions related to general stress, specific infection, trauma, joint disease and metabolic conditions increased compared to the preceding Iron Age, indicating that the arrival of the Romans had a substantial impact upon the health status of the indigenous population (Roberts and Cox, 2003: 389). Further work by Redfern and Roberts (2005) examining health in Romano-British urban centres (including York, but limited to osteological evidence from Trentholme Drive and Castle Yard) found high rates of infectious and metabolic disease, suggesting that Romano-British towns were highly unsanitary, consequently impacting upon the health of their inhabitants (Redfern and Roberts, 2005: 126).

The main aim of this study is to combine new and pre-existing osteological evidence with archaeological evidence in order to reconstruct the health status of the population of Roman York. Key research questions addressed in this paper are: what was the health status of the population? How does the health status of this population compare to similar populations residing elsewhere in Roman Britain? The study focuses on evidence for diseases and injuries related to physical activities, and on infections and other acquired pathological conditions.

2. Materials and Methods

2.1. York and Comparative Urban Population Samples

This study collated osteological data for 785 individuals of Roman date from a total of 94 archaeological interventions in York (for full details of these archaeological excavations see McIntyre, 2013). Precisely 90.4% of the York sample were adults, with the remaining 9.6% being sub-adults. The study included all skeletal material found *in situ* within 5 km of the Roman fortress, in order to encompass both the main contemporary cemetery areas located around the town areas and scattered roadside burials still associated with the town. The dearth of (typically earlier) cremation burials available for study means that the majority of burial data pertains to the mid/later part of the period of Roman occupation, from *c.* AD 150–350. Therefore, the results of this study are more likely to be informative of the population residing in the town during this time period.

One caveat to this study involves skeletal material pertaining to the 82 skeletons from sites 3 and 6 Driffield Terrace. Skeletal material from these sites was originally analysed in 2006 (Tucker, 2006), and it is this data which is included in the present study. The assemblages have since been reanalysed (Caffell and Holst, 2012): the 2012 data was not available for inclusion in the present study at the time of writing. Therefore, any new findings from the 2012 study are not incorporated into the results of

Table 8.1 Comparative osteological assemblages.

Site	Date	#Crem.	#Inhum.	Total	Reference
St. Stephens, St. Albans	Early–late Roman	297	27	324	McKinley, 1992
Southern Cemetery, London	Early 1st–late 4th century	0	46	46	MOLA Centre for Human Bioarchaeology, 2009a
Western Cemetery, London	Early 1st–early 5th century	0	137	137	MOLA Centre for Human Bioarchaeology, 2009b
120-122 London Road, Gloucester	1st–4th century	12	154	166	Simmonds *et al.*, 2008
Eastern Cemetery, London	Late 1st–4th century	136	550	686	Barber and Bowsher, 2000
Derby Racecourse, Derby	Mid 2nd–mid 4th century	39	73	112	Wheeler, 1985
Alington Avenue, Dorchester	Mid 2nd–4th century	3	91	94	Davies *et al.*, 2002
Poundbury, Dorset (Dorchester)	Late 2nd–4th century	0	1388	1388	Farwell and Molleson, 1993
Brougham	3rd century	290	0	290	Bell and Cool, 2003
Icknield Street, Dunstable	3rd–early 5th century	0	112	112	Matthews, 1981
Butt Road, Colchester	Late 3rd–late 4th century	5	728	733	Crummy *et al.*, 1993
Ancaster	Late 3rd–4th century	0	282	327	Cox, 1989
Lankhills, Winchester	4th century	32	751	783	Clarke, 1979; Booth *et al.*, 2010
Bath Gate Cemetery, Cirencester	4th–5th century	3	405	424	McWhirr *et al.*, 1982
Little Keep, Dorchester	Late Roman	0	29	29	Dinwiddy, 2009
Total		817	5073	5890	

this paper. Should the present project be reviewed in the future, it is anticipated that the 2012 data (plus data from any new discoveries) will be incorporated at that time.

Published osteological data for 5890 individuals from 11 Romano-British urban sites were chosen for comparison with the York data (Table 8.1; Figure 8.2). Comparative sites were chosen on the basis of sample size (data available for more than 100 individuals), quality and recency of data, accessibility of osteological and archaeological data for burials dating to the appropriate period of study, and the possession of significant urban settlement or military installation status during the period covered by the study. At Dorchester and London, data from more than one cemetery were combined in order to produce a larger sample and better represent the original overall population of the town.

2.2. Methods

For the skeletal samples from York, osteological data in the form of published or unpublished reports and archival notes were collected from archaeological interventions in the area of study. The majority of data produced by other researchers (and collated for this study) were gained using standard osteological methods, as outlined in Brickley and McKinley (2004). While some osteological reports (those pre-dating publication of the BABAO 2004 standards) do not necessarily conform to these exact recording standards, in many cases the same recording and/or analytical methods were in fact used to produce data. Where no osteological data (or osteological data of poor quality) were available and skeletal material was accessible, macroscopic osteological analysis was conducted by the principal author , in accordance with

recognised standard osteological guidelines (Brickley and McKinley 2004; Steckel *et al.*, 2006).

Macroscopic pathological observations were made and interpreted, initially, in accordance with Aufderheide and Rodríquez-Martín (1998) and Ortner (2003), and then with reference to a variety of relevant clinical and palaeopathological publications as part of a full differential diagnosis. Pathological lesions were recorded descriptively, and in terms of presence/absence, anatomical location and extent. Radiography, microscopic and biomolecular methods were not utilised. The principal pathological conditions recorded include trauma, congenital conditions, circulatory disorders, joint disease, infectious disease, haematological disorders, neoplastic conditions, and any miscellaneous pathology. Dental pathology was also recorded and analysed as part of the wider study, but these results are not discussed here.

True prevalence rates (TPR – the percentage of skeletal elements affected by a pathological condition) were calculated where possible. Crude prevalence rates (CPR - the percentage of individuals affected by a pathological condition) were also calculated in order to facilitate statistical comparison between York and the 11 comparative populations: raw counts of affected/ unaffected skeletal elements were unavailable in the majority of publications, meaning that calculation of TPR for comparative populations unfortunately was not usually possible. Where raw data were not available, prevalence rates were quoted from the relevant reports. Samples with zero prevalence were included in the statistical comparisons.

Chi-square tests were conducted in Microsoft Excel to test whether there was any significant difference in

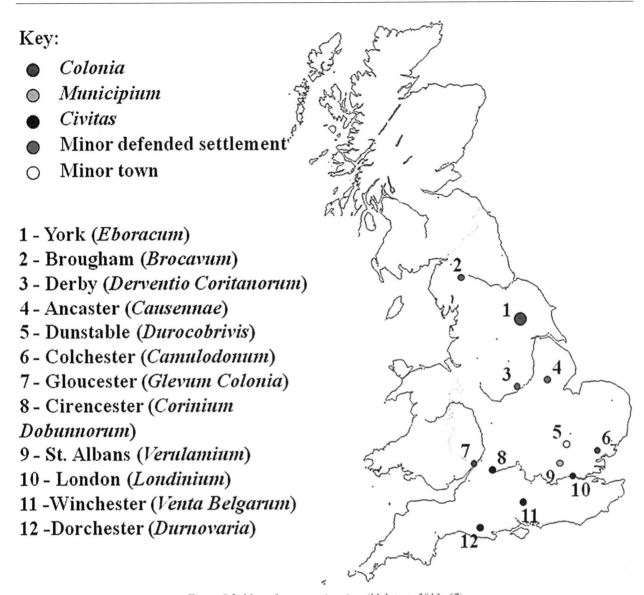

Key:

- ● *Colonia*
- ◉ *Municipium*
- ● *Civitas*
- ◉ Minor defended settlement
- ○ Minor town

1 - York (*Eboracum*)
2 - Brougham (*Brocavum*)
3 - Derby (*Derventio Coritanorum*)
4 - Ancaster (*Causennae*)
5 - Dunstable (*Durocobrivis*)
6 - Colchester (*Camulodonum*)
7 - Gloucester (*Glevum Colonia*)
8 - Cirencester (*Corinium Dobunnorum*)
9 - St. Albans (*Verulamium*)
10 - London (*Londinium*)
11 -Winchester (*Venta Belgarum*)
12 -Dorchester (*Durnovaria*)

Figure 8.2 Map of comparative sites (McIntyre, 2013: 67).

CPR between towns, and in SPSS Statistics 19 when testing between males and females from York. Where applicable, chi-square tests and Fisher Exact tests were also run in SPSS to test between manifestation of pathological conditions in the left and right sides of the body in individuals from York (i.e. in conditions that may occur preferentially on one side of the body such as os acromiale), and to test proportional survival/recovery of skeletal elements across excavated sites in York.

3.   Results

3.1.   Roman Britain

Firstly, overall pathological prevalence rates from Roman Britain (comprising data from all 12 populations including Roman York) were examined. Figure 8.3 shows the CPR distribution of trauma and activity-related conditions across all populations. Figure 8.4 shows the CPR distribution of acquired and infection-related conditions. Outlying values are not included; these figures show the

observed 'normal' distribution ranges. The coloured boxes represent the second and third quartiles, the whiskers represent the range, and the black line the median value.

Figures 8.3 and 8.4 show that across all observed populations, the largest pathological prevalence ranges are for spinal and extra-spinal degenerative joint disease and Schmorl's nodes, suggesting that prevalence of these conditions is highly variable across Romano-British urban sites. These pathological conditions (particularly spinal joint disease) also have the potential for the highest prevalence.

These figures also show that post-cranial trauma, osteochondritis dissecans, non-specific infection, and cribra orbitalia have moderate prevalence ranges at a slightly lower level (up to approximately 15% CPR), suggesting that these conditions are likely to be present in Romano-British urban populations, but with lesser frequency than the degenerative conditions. All other observed pathological conditions have consistently low and narrow prevalence ranges (0–5% CPR) indicating

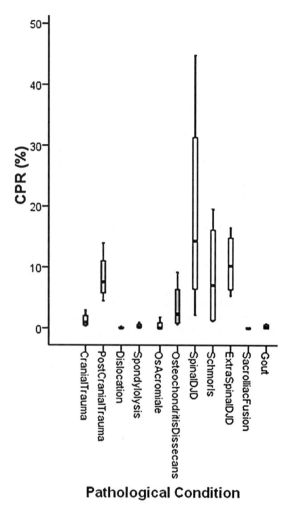

Pathological Condition

Figure 8.3 Trauma and activity-related pathology CPR at all Romano-British towns. Outlying values are not included; this Figure only shows the observed 'normal' distribution ranges. The coloured boxes represent the second and third quartiles, the whiskers represent the range, and the black line the median value.

that no more than a handful of examples of these are likely to be found in Romano-British urban populations. After establishing these patterns, it was possible to discern whether prevalence of each pathological condition found in the York population fell within or outside this distribution.

3.2. York

Prevalence rates for York combine both sub-adult and adult data. Compared to other observed Romano-British populations, York had statistically significantly elevated prevalence of ante-mortem cranio-facial and post-cranial trauma, and os acromiale. All other pathological conditions were within the ranges shown in Figures 8.1 and 8.2.

3.2.1. Trauma and Activity-Related Pathological Conditions

This section will present the observed prevalence rates and results of statistical analyses for the following pathological conditions: ante mortem fracture, dislocation, spondylolysis,

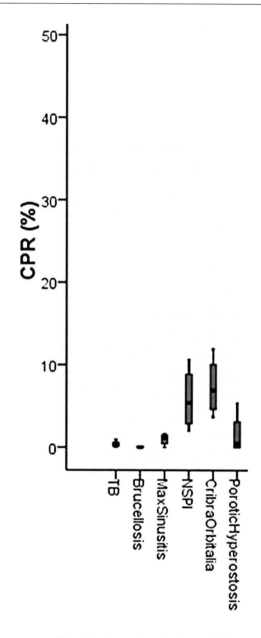

Pathological Condition

Figure 8.4 Acquired and infectious pathology CPR at all Romano-British towns. Outlying values are not included; this Figure only shows the observed 'normal' distribution ranges. The coloured boxes represent the second and third quartiles, the whiskers represent the range, and the black line the median value.

joint disease, Schmorl's nodes, osteochondritis dissecans and os acromiale.

Twenty-two individuals in the York assemblage were observed with evidence of ante-mortem cranio-facial trauma (CPR = 2.8%, also Table 8.2). Cranio-facial trauma occurred at a significantly higher rate in males than in females (Table 8.2). Table 8.2 also shows that many of the individuals affected by cranio-facial trauma are from burial sites along Driffield Terrace, and there was a significantly higher prevalence of cranio-facial trauma at the Driffield Terrace locations compared to other locations in York.

Table 8.2 CPR of cranio-facial trauma at York, comparative towns and Roman Britain.

	Total no. individuals	Total affected	M	F	M CPR%	F CPR%	Total CPR%	Reference
York (overall)	785	22	21	1	7.14	0.64	2.80	This study
York (Driffield Terrace)	82	16	16	0	25.4	0.00	19.51	This study
York (all other sites)	703	6	4	1	1.73	0.64	0.85	This study
Ancaster	327	1					0.31	Cox, 1989
Cirencester	424	5	5	0	2.04	0.00	1.18	McWhirr et al., 1982
Colchester	733	5	1	3	0.55	1.69	0.68	Crummy et al., 1993
Derby	112	0					0.00	Wheeler, 1985
Dorchester	1511	12	8	2	1.82	0.43	0.79	Farwell and Molleson, 1993; Davies et al., 2002; Egging Dinwiddy, 2009
Dunstable	112	0					0.00	Matthews, 1981
Gloucester	166	2	2	0	3.23	0.00	1.20	Simmonds et al., 2008
East London	686	>1					0.15	Barber and Bowsher, 2000
Winchester	783	10	7	2	3.27	0.92	1.28	Booth et al., 2010
Roman Britain	2083	24					1.15	Roberts and Cox, 2003

Table 8.3 CPR of post-cranial trauma at York and comparative towns.

	Total no. individuals	Total affected individuals	M	F	M CPR%	F CPR%	Total CPR%	Reference
York	785	63	49	5	16.67	3.21	8.03	This study
York (Driffield Terrace)	82	38	34	0	53.97	0.00	46.34	This study
York (all other sites)	703	25	16	5	6.93	3.23	3.56	This study
Ancaster	327	23	17	5	13.18	6.02	7.03	Cox, 1989
Cirencester	424	59	51	7	20.82	7.00	13.92	McWhirr et al., 1982
Colchester	733	33	15	10	8.29	5.65	4.50	Crummy et al., 1993
Derby	112	2	2	0	10.53	0.00	1.79	Wheeler, 1985
Dorchester	1511	57	34	12	7.73	2.58	3.77	Farwell and Molleson, 1993; Davies et al., 2002; Egging Dinwiddy, 2009
Dunstable	112	7	5	1	10.87	2.38	6.25	Matthews, 1981
Gloucester	166	16	10	0	16.13	0.00	9.64	Simmonds et al., 2008
Winchester	783	33	18	10			4.21	Booth et al., 2010

Sixty-four individuals provided evidence for a total of 112 traumatic injuries (CPR=8.0%: Table 8.3). Affected individuals had between one and four traumatic injuries to the post-cranial skeleton. Crude post-cranial trauma prevalence in the above urban populations ranged between 1.79% and 13.92%, with the differences between samples being statistically significant (χ^2=82.52, p<0.01). With a CPR of 8.0%, the York assemblage sits in the middle of this distribution (also see Figure 8.3). The York males had significantly more post-cranial traumatic injuries than females.

Over half the individuals with post-cranial trauma from York were from burial locations at Driffield Terrace

(38/64 individuals). The chi-square value (χ^2=151.87, p=<0.01), indicates that there is a substantial difference in prevalence of post-cranial trauma between Driffield Terrace and all other sites.

Skeletal evidence for two (possible) dislocations were found in the York assemblage, both in adult male individuals (CPR=0.25%). One dislocation was found in the right elbow joint (1/656 joints, TPR 0.31%) and the other in the left hip (1/384 joints, TPR 0.53%). These two cases are unlikely to represent the prevalence rates for all dislocated joints in the York population, only those that were not treated or not treated successfully. Of all observed populations, only three (York, Dorchester and Gloucester)

Table 8.4 TPR of os acromiale per affected skeletal element.

Element	Total elements	Total affected elements	L	R	L TPR%	R TPR%	Total TPR%
Scapula	311	17	11	6	7.59	3.95	5.47

had examples of dislocated joints, corroborating the low prevalence of skeletal evidence for dislocation previously reported for Roman Britain (CPR=0.46%: Roberts and Cox, 2003: 158).

Seven individuals were observed with spondylolysis, in either the fourth (1/152 vertebrae, TPR=0.66%) or fifth lumbar vertebrae (5/152 vertebrae, TPR=3.29%). In one individual, the exact lumbar vertebra affected was unknown; this individual can therefore only be included in crude prevalence calculations. CPR of spondylolysis in the York assemblage was 0.89%. There was no significant difference in prevalence of spondylolysis between males and females.

A total of 98 individuals were observed with spinal joint disease (CPR=12.5%). No significant difference was found between the sexes. The observed prevalence rate of spinal joint disease at York is close to the median CPR for Roman Britain (Figure 8.3), across all populations, York had the third highest prevalence of spinal joint disease, after Cirencester and St. Albans. Differences in prevalence across the comparative sites were statistically significant (Fisher's Exact=387.74, p=<0.01),

A total of 103 individuals had extra-spinal joint disease (CPR=13.1%). York sits at the higher end of the prevalence range when observing all comparative populations (Figure 8.3), and variation amongst the comparative samples was statistically significant (Fisher's Exact=117.07, p=<0.01). No significant difference was found between males and females in the York assemblage, which is consistent with previously published findings (Aufderheide and Rodríguez-Martín, 1998: 93).

A total of 99 individuals from the York assemblage had Schmorl's nodes. (CPR=12.6%). CPR at York is higher than the median value for the overall prevalence range. The differences between the comparative samples are statistically significant (Fisher's Exact=388.20, p=<0.01). Males had a significantly higher prevalence of Schmorl's nodes than females (χ^2=23.10, p=<0.01). This is consistent with previous research regarding sex distribution of Schmorl's nodes in skeletal samples (Faccia and Williams, 2008: 36; Dar *et al.*, 2009).

Osteochondritis dissecans was observed in eight individuals (CPR=1.0%), and only in bones of the lower legs and feet. Roberts and Cox (2003: 151) report an average of 0.2% in Roman British skeletal samples. In view of this, data values for osteochondritis dissecans appear rather large in Figure 8.3. However, this range is likely to have been markedly affected by the higher rates observed at Cirencester and (to a lesser extent), Colchester and St. Albans. There was significant variation in the prevalence of osteochondritis dissecans across the comparative samples (Fisher's Exact=127.14, p=<0.01), although this again may be attributable to the high

frequency at Colchester. CPR at York was lower than the median.

Os acromiale was observed in 14 individuals from York (CPR=1.8%; of these 14 only three had bilateral involvement. Table 8.4 shows TPR of os acromiale in the York assemblage (n. b. 145 left scapulae, 152 right scapulae, and 14 unsided). The prevalence of os acromiale was significantly higher in the left compared to the right side (χ^2=76.246, p=<0.001). If this finding is reflective of occupational activity, it indicates that repeated mechanical stress was more likely to be placed on the left shoulder then the right. York was the only site in this study where os acromiale was observed. It is unknown whether this is due to genuine absence or differential recording strategy (at least in some of the older reports) which resulted in the condition not being recorded. The previously calculated prevalence for Roman Britain utilises a much smaller sample than that of York (10/308 individuals from four sites, giving a CPR of 3.2%, (Roberts and Cox, 2004: 158, though when sites with zero prevalence are included the value falls to 0.2%). Os acromiale was also present significantly more frequently in males than females in the York assemblage (χ^2=4.30, p=0.04). Both the sex difference and the side bias strongly supports a mechanical aetiology over genetic predisposition for the condition within the York population (Case *et al.*, 2006: 13).

3.2.2. Acquired and Infection-Related Pathological Conditions

This section will present the observed prevalence rates and results of statistical analyses for the following pathological conditions: tuberculosis, brucellosis, maxillary sinusitis, non-specific infection (combined incidence of periostitis, osteitis and osteomyelitis), cribra orbitalia and porotic hyperostosis.

Two male adults from York had possible gastro-intestinal tuberculosis. CPR of tuberculosis at York was very low (0.25%); the prevalence of tuberculosis across all the observed comparative sites is between 0–1%, and therefore also relatively low, corroborating a previous low estimate of CPR=0.2% in Roman Britain (Roberts and Cox, 2003: 119).

One case of possible brucellosis was observed in an older adult female (Holst, 2010). The only skeletal elements involved were both ilia of the pelvis, giving a TPR for Roman York of 0.52%. Post-depositional damage to the affected bones meant that this case could not be confirmed (Holst, 2010: 4). Alternative proposed diagnoses include tuberculosis and osteomyelitis (Holst, 2010: 6). No other cases of brucellosis (possible or confirmed) were found in any of the comparative sites used in this study, and the condition has rarely been seen in British skeletal samples (Roberts and Cox, 2003: 230).

Table 8.5 CPR of porotic hyperostosis at York and comparative towns.

	Total no. individuals	Total affected	M	F	M CPR%	F CPR%	Total CPR%	Reference
York	785	42	27	9	9.18	5.77	5.35	This study
Ancaster	327	3	1	0	0.78	0.00	0.92	Cox, 1989
Cirencester	424	0					0.00	McWhirr et al., 1982
Colchester	733	0					0.00	Crummy et al., 1993
Derby	112	0					0.00	Wheeler, 1985
Dorchester	1511	0					0.00	Farwell and Molleson, 1993; Davies et al., 2002; Egging Dinwiddy, 2009
Dunstable	112	0					0.00	Matthews, 1981
Gloucester	166	0					0.00	Simmonds et al., 2008
London	869	1	0	1	0.00	0.56	0.12	Barber and Bowsher, 2000; MOLA Centre for Human Bioarchaeology, 2009a, 2009b
Winchester	783	0					0.00	Booth et al., 2010

Seven individuals from York were observed with maxillary sinusitis (CPR=0.9%). The prevalence of maxillary sinusitis is similar to the prevalence of 0.6% established for a large sample of 5716 individuals in Roman Britain (Roberts and Cox, 2003: 112) and for the comparative samples in this study. Additionally, no significant difference was found in the distribution of maxillary sinusitis at York according to sex.

Compared to the other observed populations, York has the third highest prevalence of non-specific infection (all categories of non-specific infection were combined in CPR calculations). There were significant differences in prevalence amongst the comparative samples (Fisher's Exact=258.05, p=<0.01). In the York sample significantly more males than females were affected by non-specific infection (χ^2=4.74, p=0.03).

In the comparative sample there were significant differences in prevalence of cribra orbitalia (Fisher's Exact=272.09, p=<0.01). Values observed at York (44/785, CPR 5.6%), are at the lower end of the expected prevalence range (Figure 8.4). No significant difference was found in prevalence of cribra orbitalia according to sex or side in the York population.

Observed prevalence of porotic hyperostosis at York was more than five times higher than in the comparative populations (42/785, CPR=5.4%, also see Table 8.5). Porotic hyperostosis was only observed in two of the comparative populations (Ancaster, 3/327, CPR 0.9%, and London, 1/869, CPR=0.1%). Differences amongst the comparative samples were significant (Fisher's Exact=139.87, p=<0.01).

4.   Discussion – Health in *Eboracum*

The results of this study show that when compared to frequencies found at other Romano-British urban sites, the majority of pathological conditions observed in the population from York fall within the expected (crude prevalence) range for the period. In terms of CPR, the population had noticeably elevated prevalence of os acromiale, cranial trauma, post-cranial trauma. Prevalence of trauma, os acromiale, Schmorl's nodes, and non-specific infection were also observed as being significantly different between males and females from York.

4.1.   Trauma, Physical Stress and Occupational Activity

Prevalence of os acromiale appears to be elevated in the York skeletal sample compared to the other comparative urban Roman cemetery sites, none of which had recorded incidences of os acromiale. Prevalence of os acromiale varies between modern populations (Case et al., 2006), with the condition occurring in approximately 3–8% of individuals (Miles, 1994: 150). These modern clinical observations mainly refer to North American or European populations, and this prevalence range does appear to corroborate results obtained from archaeological studies utilising European and North American archaeological and anthropological specimens (Miles, 1994; Case et al., 2006; Hunt and Bullen, 2007). Moreover, if British and European populations are likely to have a prevalence range of 3–8%, this may suggest that archaeologically observed rates from Roman Britain are lower than expected. Alternatively, os acromiale may have been under-reported in unpublished and published osteological reports, as some authors have classified the condition as a non-metric trait rather than as a pathological lesion (Connell and Rauxloh, 2003: 13; Hunt and Bullen, 2007: 316) and this may have contributed to the under-reporting of os acromiale in palaeopathological reports.

The observed sex and side bias in the occurrence of os acromiale strongly supports a mechanical aetiology over genetic predisposition for the condition within the Roman population of York, Therefore, it is suggested that males from York were more likely to be engaging in activities that placed increased stress on the rotator cuff, which in turn increased their likelihood of impinged acromial fusion.

Development of os acromiale is difficult to relate to a particular activity, and the trait is not occupation specific, though high prevalence has been reported in military cohorts, e.g. the crew of the Mary Rose (Stirland, 2005), the mass grave at Towton (Coughlan and Holst, 2000), and early medieval cemeteries in Ireland (Geber, 2015). In these cases the defect has been linked to participation in strenuous activity from an early age. Any activity placing stress on the rotator cuff muscles of the shoulder, especially in late adolescence, may contribute to the development of the condition (Coughlan and Holst, 2000: 73). In modern patients, os acromiale is often found in athletes, with clinically documented cases often involving basketball and baseball players, and American footballers (Paley *et al.*, 2000; Pagnani *et al.*, 2006). Presumably individuals who develop os acromiale are more likely to be partaking in such activities from at least adolescence, if not before: the acromiale epiphysis commences fusion at approximately age 17–18 and should be fully fused in the majority of people by age 20 (Case *et al.*, 2006; Schaefer *et al.*, 2009: 164). This evidence therefore indicates that males at *Eboracum* are more likely to be partaking in physical activities during adolescence that may impinge acromiale fusion.

A second pathological lesion with links to activity patterns was also found significantly more frequently in males than females: Schmorl's nodes. Mechanical loading may exacerbate the presence and location of Schmorl's nodes, and torsional movement, in particular, is thought to be a major causal factor (Dar *et al.*, 2010: 673).

Presence of os acromiale and Schmorl's nodes appears to be related to physical stress. The men residing in Roman York have been associated with a variety of occupations, relating to the military, industry, crafts, politics, trade and so on (Ottaway, 2004: 59–61). However, there is very little evidence pertaining to female activity during the same period: epigraphic references to women are mostly associated with women of higher social status, and tend to focus on their role as a wife and mother (Ottaway, 2004: 117–120). In terms of wider Roman (or Romanised) society, literary sources frequently discuss male and female roles in terms of the Roman 'ideal', that is that men are involved with politics, administration, and the military, and that women are involved with the domestic side of life; being a wife and mother, and looking after the household (Scheidel, 1995: 205–206; Hemelrijk, 2004; Revell, 2005). In reality, these idealised roles are more likely to have been adhered to by the wealthier classes (Scheidel, 1995: 207). For everyone else, these idealised roles are much less viable, and presumably both men and women would be found undertaking a wide variety of occupations. A cursory scan of ancient literary sources and modern texts discussing female occupations could lead to the belief that the majority of women not lucky enough to marry a rich man would have been sold as slaves or worked in brothels (Clark, 1981: 197-8). While this was

undoubtedly the case for some women, and women would also be the most likely candidates to take care of domestic chores and rearing children, those with jobs and careers are equally likely to have been working in industry (e.g. textiles), agriculture, trade and production (Clarke, 1981: 197–198; Allason-Jones, 1989: 64, 70, 80–1; Scheidel, 1995: 205–7).

Significantly higher prevalence of both cranio-facial and post-cranial trauma was observed at York compared to the other observed towns. Cranio-facial trauma was found in the form of healed nasal and maxillary fractures, and depressed cranial fractures. These types of injury are most commonly caused by interpersonal violence (e.g. being hit in the face) and accidental/intentional injury by low velocity blunt force trauma by small projectiles (e.g. slingshot). High levels of cranio-facial trauma and injury, therefore, suggest an elevated level of interpersonal violence at York compared to other urban settlements (although injury as a result of a deliberate violent act does not imply active involvement by the recipient: Brickley and Smith, 2006: 171). Facial trauma, in particular, is most often associated with assault (Walker, 2001: 582–583; Alvi *et al.*, 2003: 103; Lee *et al.*, 2010: 194); unlike post-cranial trauma, accidental injury is less likely to occur to the head and face (Webb, 2009: 202).

Post-cranial trauma was observed in all regions of the body, and was fairly evenly distributed in terms of location prevalence. Although trauma rates per post-cranial skeletal element were consistently low (with the exception of the intermediate and distal phalanges of the feet), overall frequency and crude prevalence was still significantly higher than at all other observed comparative sites, and particularly high amongst the Driffield Terrace burials. The observed post-cranial injuries probably result from a combination of deliberate and accidental traumatic events. Patterns of injury are highly culturally specific, being likely to vary both between and within populations (Brickley and Smith, 2006: 165–166). Types of violence are also culture specific, particularly in terms of factors such as weapon choice, or even social class (Walker, 2001; Brickley and Smith, 2006: 168–170).

Statistically, males had significantly higher prevalence of both cranio-facial and post-cranial trauma than females. Sex distribution of cranio-facial fracture has been found to vary greatly from study to study (both clinical and osteological), with biases towards either sex or equal distribution between the sexes (Shepherd *et al.*, 1988; Shepherd *et al.*, 1990: 77; Walker, 2001: 582; Gassner *et al.*, 2003: 53; Judd, 2004). This suggests that distribution of cranio-facial trauma between the sexes is likely to be influenced by socio-cultural factors associated with the population.

Individuals buried at sites located Driffield Terrace, a well-known Roman cemetery area located on the south west side of York, were more likely to have experienced cranial and post-cranial trauma. All individuals with traumatic lesions from sites located on Driffield Terrace were males aged approximately 17–45 years. The combination of the male-dominated assemblage at sites on and immediately adjacent to Driffield Terrace (e.g. 129 The Mount, sites at The Mount School), plus a preponderance of decapitation burials, has been used as evidence to suggest that this area was used as a gladiator cemetery (Hunter-Mann, 2006a, 2006b; Müldner *et al.*, 2011: 9). Alternative explanations have suggested the graves are those of executed criminals, that the area was used as a military cemetery, or that these individuals were members of a religious group or cult (Hunter-Mann, 2006a; Müldner *et al.*, 2011: 9; York Archaeological Trust, 2011). Montgomery *et al.* (2011) noted that the Driffield Terrace group were unlikely to be 'common criminals', because of the high-status burial location on the main road. Other accounts of Roman decapitation burials have associated this funerary rite with high or normative status rather than with punishment (Harman *et al.*, 1981; Philpott, 1991; Anderson, 2001).

What the present study can confirm is that this group of individuals are likely to represent a male-dominated group, such as a burial club, social, occupational or community group, with a predisposition towards physical, traumatic injury (accidental or otherwise), diverse geographical origin and differential burial rites. This may include, but is not limited to a military group or gladiatorial school. The presence of individuals with high frequency of trauma elevates the prevalence of trauma for the town as a whole compared to other Roman urban cemeteries in Britain.

4.2. Developmental Health and Acquired Diseases

Higher prevalence of infectious disease (specific or non-specific) is often attributed to factors such as high or increased population size and density, transition to permanent housing and/or sedentary lifestyle, increased interregional mobility and contact, poor sanitation, and insufficient nutritional uptake (Wood *et al.*, 1992: 358; Grauer, 1993: 204; Manchester, 1995: 10; Roberts, 2000: 147; Peck, 2009: 147). When considering the development of York and changes in settlement type instigated by the arrival of the Romans in the area, it is likely that all these factors were at play, and would have contributed to disease levels and susceptibility of the population to infection. Peck's (2009: 114) study shows that the Romano-British population from Trentholme Drive had five times the frequency of periostitis compared to the Iron Age population from East Yorkshire (when comparing mixed sex prevalence rates).

Furthermore, the arrival of the Roman army, establishment of the fortress and eventual development of the town meant that for the first time large numbers of people were living in much closer quarters. Higher population density may have served to promote the transmission of infection.

The picture regarding town sanitation in Roman York is rather mixed. On the one hand, the fortress buildings appear to be well maintained, with good drainage, a clean water supply, and regular refurbishment (Addyman,

1989: 246; Ottaway, 2004: 67–75, 94). The construction of a large stone sewer at Church Street indicates that measures were being taken to remove human waste from living areas (Buckland, 1974; Ottaway, 2004: 44–46). The presence of several public and military bathhouses (examples of public baths have been excavated at Micklegate, Toft Green, Bishophill, the Old Station and Fetter Lane, and military bathhouses are located within the fortress: York Archaeological Trust, 1973; Ottaway, 2004: 42, 92; Robinson, 2012, pers. comm.) indicates that that provisions were also made for public bathing and personal cleanliness. However, the regularity with which such public facilities would be cleaned is debateable: bathhouses may, therefore, have aided in the transmission of bacteria and infection rather than facilitating good personal hygiene (Allason-Jones, 1989: 83).

Conversely, environmental evidence from several sites immediately adjacent to the River Ouse suggests that these areas were used as dumping grounds for general waste in the earlier periods of occupation, to the extent that these areas may have posed a health hazard (Addyman, 1989: 250). Furthermore, faunal evidence suggests that vermin such as rats and mice were a common problem in the town from the 2nd century (Addyman, 1989: 250; Ottaway, 2004: 108). Evidence for the presence of vermin and dumping of human and general waste has also been found at the General Accident site: these activities are likely to date to the 2nd century, but it is unclear whether this would have been a typical phenomenon throughout York at this time, or whether this is merely an example of localised rubbish deposition (York Archaeological Trust, 1984; Ottaway, 2004: 108–109). Analysis of waste deposits from the Church Street sewer has also indicated the presence of human gut parasites, and mid-2nd century drain and ditch fills from Tanner Row and Rougier Street have contained the eggs of two species of intestinal parasitic worms (*Trichuris* and *Ascaris*, though both species may be carried by animals and do not prove the sewer contained human excrement: Buckland, 1974; Wilson and Rackham, 1976: 32; Dobney *et al.*, 1999: 20). Increased rubbish dumping and presence of 'dark earth' from the 4th century indicates that standards of sanitation and hygiene in the town are likely to have declined in the later period, which is likely to have increased the population's susceptibility to infection and ill health (York Archaeological Trust, 1973, 1981, 1984, 1987; Ottaway, 1999: 147; Faulkner, 2001: 124; Ottaway, 2004: 146, 149).

Rates of infection in the York population show significant male bias. This is especially interesting when also taking into consideration Peck's (2009: 148) finding that Romano-British males from Trentholme Drive were at higher risk of periostitis than Iron Age males from East Yorkshire, whereas females from the same populations were at approximately the same risk. Frequency of non-specific infection was also higher in males from eight out of nine observed comparative towns. Although these rates could not be tested statistically to see if the observed bias was significant (because of insufficient raw comparative data), male biased infection rates do appear to be repeated across Romano-British urban populations. Generally, male morbidity for infection is greater, although other factors such as increased female immunity or males being exposed to more infectious agents may also be contributing factors (Ortner, 1998: 80–86). However, ratios are also likely to be affected by time period, geographical region, socio-cultural and other factors, meaning that the precise cause of the observed sex bias in non-specific infection prevalence is likely to be complex and multi-aetiological.

Porotic hyperostosis and cribra orbitalia have been associated with iron deficiency anaemia and also Vitamin B deficiency, macroscopic observation of these lesions is best used as being evidence of general stress (Steckel *et al.*, 2006: 12–13). Results show that prevalence of cribra orbitalia appears to vary substantially between all observed populations, with rates being at the lower end of the expected range at York. Prevalence of porotic hyperostosis follows a different pattern, being consistently low at comparative sites, but with anomalously high prevalence at York. It is the opinion of the authors that observed rates of porotic hyperostosis at York more closely approximate genuine prevalence of the condition in the living population, with heavily remodelled lesions perhaps being overlooked osteologically.

This may also be evident when comparing CPR of cribra orbitalia and porotic hyperostosis in the same population: frequency and prevalence rates of these conditions at York are very similar, where at all other comparative sites, porotic hyperostosis rates are much lower. This may be a reflection of the fact that cribra orbitalia lesions, even when substantially remodelled, are easy to recognise osteologically. The somewhat more subtle appearance of remodelled porotic hyperostosis lesions can render them easy to miss. Furthermore, differential recording strategies employed by osteologists may lead to differential diagnosis of more lesions that are more subtle in appearance (Jacobi and Danforth, 2002). Significantly lower prevalence of porotic hyperostosis in the comparative samples therefore likely pertains to observational difficulty or differential diagnosis within archaeological samples. Cribra orbitalia prevalence at York is likely to be genuine, and hence, low for the period.

Considering that frequency and crude prevalence of cribra orbitalia and porotic hyperostosis is very similar at York, these figures may in fact reflect relatively low overall prevalence for both conditions. Therefore a tentative explanation for the observed results may indicate that the population was more likely to be subject to dietary (and other) stressors during childhood, with lower/moderate exposure to stressors during adolescence and adulthood. The latter theory is substantiated by recent research by Pitts and Griffin (2012), whose study of late Romano-British cemetery sites found that health and nutrition was significantly better at urban and nucleated sites when compared to their rural counterparts. Patterns of health were found to be strongly associated with

site type (i.e. dependant on whether the site was urban, nucleated or rural) as well as road connectivity and diet (Pitts and Griffin, 2012: 260, 272). This may indicate that the health status and diet at York was better than at nearby rural sites.

At York, general patterns of health generally conform to the trends observed in populations from all towns. It should be noted that patterns of health at York are more likely to relate to the later period of occupation, because of the paucity of skeletal material relating to the earlier period. Generally, adult male individuals from the town are more likely to be physically active, partaking in activities that place stress on the left shoulder joint or increase torsional rotation of the spine. Probably the most stand out pathological results relate to trauma, with individuals from York being more likely to experience traumatic injury than individuals from the comparative populations.

On the whole, the health status of the population of Roman York is comparable to that observed at similar, contemporary urban sites. General levels of stress within the population appear to be low, though levels of activity related mechanical stress and non-specific infection are higher. Higher levels of traumatic injury (particularly pertaining to the cranium) are predominantly found in an unusual group of male burials located around the Driffield Terrace area of the main cemetery on the south west road heading out of the town. Therefore, trauma prevalence in the rest of the population falls within the expected (low) prevalence range.

5. Conclusion

Collation and analysis of osteological data from 94 sites from Roman York, and a further 11 contemporary towns has facilitated reconstruction of broad patterns of health in across Romano-British urban centres. Typically, populations from Romano-British urban sites are likely to have higher and more variable rates of degenerative joint disease and Schmorl's nodes, moderate levels and range of conditions such as post-cranial trauma, osteochondritis dissecans, non-specific infection, and cribra orbitalia, and low levels and narrow ranges of all other pathological conditions.

On the whole, the health status of the population of Roman York is comparable to the health status observed at similar, contemporary urban sites. It should be noted that patterns of health at York are more likely to relate to the later period of occupation, because of the paucity of skeletal material relating to the earlier period. Generally, adult male individuals from the town are more likely to be partaking in physical activities that place stress on the left shoulder joint or increase torsional rotation of the spine. Males were also more prone to non-specific infection, and this pattern appeared to be repeated across most contemporary towns observed in this study.

Probably the most stand out pathological results relate to trauma, with individuals from York being more likely to experience traumatic injury than individuals from the comparative populations. A group of males buried in the location now known as Driffield Terrace were significantly more prone to traumatic injury in the cranium and several locations in the post-cranial skeleton than the rest of the population, and, indeed, populations from other urban Romano-British sites. This group is highly likely to represent a distinctive male-dominated, traumatic injury-prone community residing within the town.

Acknowledgements

Thanks go to the Arts and Humanities Research Council for funding this study as part of a doctoral thesis at the University of Sheffield.

Thanks are also owed to a great deal of people, companies and institutions for access to archives, skeletal material, unpublished reports and osteological data: Stephany Leach at Bournemouth University, Percival Turnbull at Brigantia Archaeological Practice, Adrian Gollop at Canterbury Archaeological Trust, Charlotte Roberts at the University of Durham, Field Archaeology Specialists, Harrogate Museums and Arts, Joshua Peck at JPAC (Hawaii), Leeds Museum and Galleries, MAP Archaeology, Mike Griffiths and Associates, Fraser Hunter and The National Museum of Scotland, The Natural History Museum (London), Northern Archaeological Associates, Louisa Matthews and the North Yorkshire Historic Environment Record, On Site Archaeology, Magdalena Wachnik at Oxford Archaeology, Patrick Ottaway at PJO Archaeology, Gündula Müldner at the University of Reading, Carrie Sulosky-Weaver at the University of Virginia, Jacqueline McKinley at Wessex Archaeology Ltd., West Yorkshire Archaeology Services, Katie Tucker at the University of Winchester, John Oxley at the City of York Sites and Monuments Record, Don Brothwell, Cath Neal and the University of York, Kurt Hunter-Mann, Peter Connelly and York Archaeological Trust, Andrew Morrison and Natalie McCaul at York Museums Trust (Yorkshire Museum), and Malin Holst at York Osteoarchaeology Ltd.

References

Addyman, P.V. 1984. York in its Archaeological Setting, in: Addyman, P.V., Black, V.E. (Eds.), *Archaeological Papers from York Presented to M. W. Barley*. York Archaeological Trust, York, pp. 7–21.

Addyman, P.V. 1989. The archaeology of public health at York, England. *World Archaeology* 21 (2), 244–264.

Allason-Jones, L. 1989. *Women in Roman Britain*. British Museums Publications Ltd., London.

Alvi, A., Doherty, T., Lewen, G. 2003. Facial fractures and non-concomitant injuries in trauma patients. *The Laryngoscope* 113 (1), 102–106.

Anderson, T. 2001. Two decapitations from Roman Towcester. *International Journal of Osteoarchaeology* 11 (6), 400–405.

Aufderheide, A.C., Rodríguez-Martín, C. 1998. *The Cambridge Encyclopaedia of Human Paleopathology*. Cambridge University Press, Cambridge.

Barber, B., Bowsher, D. 2000. *The Eastern Cemetery of Roman London. Excavations 1983–1990*. Museum of London Archaeology Service Monograph 4. Museum of London Archaeology Service, London.

Bell, M., Cool, H. 2003. *The Roman Cemetery at Brougham, Cumbria Excavations 1966-67.* Description of the project and the dataset for the ADS. Available at: http://archaeologydataservice. ac.uk/archives/view/brougham_2003/downloads. cfm?CFID=5135110andCFTOKEN=78190419and (accessed 28th May, 2012).

Bendry, R., Taylor, G. M., Bouwman, A.S., Cassidy, J.P. 2008. Suspected bacterial disease in two archaeological horse skeletons from southern England: palaeopathological and biomolecular studies. *Journal of Archaeological Science* 35 (6), 1581–1590.

Blank, F. 2000. Milk borne diseases: An historic overview and status report, in: Walker, H. (Ed.), *Milk: Beyond the Dairy*. Proceedings of the Oxford Symposium on Food and Cooking, 1999. Oxford Symposium, Oxford pp. 81–85.

Booth, P., Simmonds, A., Boyle, A., Clough, S., Cool, H.E.M., Poore, D. 2010. *The Late Roman Cemetery at Lankhills, Winchester: Excavations 2000–2005*. Oxford Archaeology Monograph No. 10. Oxford Archaeology Ltd, Oxford.

Brickley, M., McKinley, J.I. 2004. *Guidelines to the Standards for Recording Human Remains*. Available at: http://www. babao.org.uk (accessed 21st July, 2011).

Brickley, M., Smith, M. 2006. Culturally determined patterns of violence: Biological anthropological investigations at a historic urban centre. *American Anthropology* 108 (1), 163–177.

Brinklow, D.A. 1984. Roman settlement around the legionary fortress at York, in: Addyman P.V., Black V.E. (Eds.), *Archaeological Papers from York Presented to M. W. Barley*. York Archaeological Trust, York, pp. 22–27.

Buckland, P. 1974. Archaeology and environment in York. *Journal of Archaeological Science* 1 (4), 303–316.

Caffell, A., Holst, M. 2012. *Osteological analysis, 3 and 6 Driffield Terrace, York, North Yorkshire: Site Codes: YORYM:2004.354 and YORYM:2005.513*. Unpublished report no. 0212, prepared for York Archaeological Trust.

Canci, A., Nencioni, L., Minozzi, S., Catolano, P., Caramella, D., Fornaciari, F. 2005. A case of healing spinal infection from Classical Rome. *International Journal of Osteoarchaeology* 15 (2), 77–83.

Capasso, L. 1999. Brucellosis at Herculaneum (79AD). *International Journal of Osteoarchaeology* 9 (5), 277–288.

Case, D. T., Burnett, S. E., Nielsen, T. 2006. Os acromiale: Population differences and their etiological significance. *HOMO – Journal of Comparative Human Biology* 57 (1), 1–18.

Clarke, G. 1979. *Winchester Studies 3; Pre-Roman and Roman Winchester. The Roman Cemetery at Lankhills*. Clarenden Press, Oxford.

Clarke, G. 1981. Roman women. *Greece and Rome* 28 (2), 193–212.

Connell, B., Rauxloh, P. 2003. *A rapid method for recording human skeletal data*. Museum of London Unpublished Report, London.

Corbel, M.J. 1997. Brucellosis: An overview. *Emerging Infectious Diseases* 3 (2), 213–221.

Coughlan, J., Holst, M. 2000. Health status, in: Fiorato, V., Boylston A., Knüsel C. (Eds.), *Blood Red Roses – The Archaeology of a Mass Grave from the Battle of Towton AD 1461*. Oxbow Books, Oxford, pp. 60–76.

Cox, M. 1989. *The human bones from Ancaster*. English Heritage, Ancient Monuments Laboratory Report 93/89.

Crummy, N., Crummy, P., Crossan, C. 1993. *Excavations of Roman and later cemeteries, churches and monastic sites in Colchester, 1971–88*. Colchester Archaeological Report 9. Colchester Archaeological Trust, Colchester.

Dar, G., Peleg, S., Masharawi, Y., Steinberg, N., May, H., Hershkovitz, I. 2009. Demographical aspects of Schmorl's nodes: A skeletal study. *Spine* 34 (9), 312–315.

Dar, G., Masharawi, Y., Peleg, S., Steinberg, N., May, H., Medlej, B., Peled, N., Hershkovitz, I. 2010. Schmorl's nodes distribution in the human spine and its possible etiology. *European Spine Journal* 19 (4), 670–675.

Davies, S.M., Bellamy, P.S., Heaton, M.J., Woodward, P.J. 2002. *Excavations at Alington Avenue, Fordington, Dorchester, Dorset*. Dorset Natural History and Archaeological Society Monograph Series 15.

De la Bédoyère, G. 2003. *Roman Towns in Britain*. Batsford/ English Heritage, London.

Dinwiddy, K. 2009. *A late Roman cemetery at Little Keep, Dorchester, Dorset*. Report for Wessex Archaeology, available at: http://www.wessexarch.co.uk/system/files/Little_Keep_ Dorchester_64913.pdf (accessed 13th July, 2011).

Dobney, K., Hall, A., Kenward, H. 1999. It's all garbage...A review of bioarchaeology in the four English colonia towns, in: Hurst, H. (Ed.), *The Coloniae of Roman Britain: New Studies and a Review*. Papers of the Conference held at Gloucester on 5–6 July, 1997. Journal of Roman Archaeology Supplementary Series No. 36, pp. 15–36.

Faccia, K.J., Williams, R.C. 2008. Schmorl's nodes: Clinical significance and implications for the bioarchaeological record. *International Journal of Osteoarchaeology* 18 (1), 28–44.

Farwell, D.E., Molleson, T.I. 1993. *Poundbury*, Volume 2: The Cemeteries. Dorset Natural History and Archaeological Society. Monograph Series Number 11.

Faulkner, N. 2001. *The Decline and Fall of Roman Britain*. Tempus Publishing Ltd., Stroud.

Field, A. 2007. *Discovering Statistics using SPSS*. Sage Publications Ltd., Thousand Oaks.

Field Archaeology Specialists. 2003. *2–4 Driffield Terrace, The Mount, York; Archaeological Evaluation (Site YDT03)*. Unpublished report.

Field Archaeology Specialists. 2005. *Archaeological excavation and watching brief; Blue Bridge Lane and Fishergate House, York (Site YBB and YFH 00-04)*. Unpublished report.

Field Archaeology Specialists. 2006. *Moss Street Depot, York; Archaeological excavation (Site YMD04)*. Unpublished report.

Gassner, R., Tuli, T., Hächl, O., Rudisch, A., Ulmer, H. 2003. Cranio-maxillofacial trauma: A 10 year review of 9543 cases with 21,067 injuries. *Journal of Cranio-Maxillofacial Surgery* 31 (1), 51–61.

Geber, J. 2015. Comparative study of perimortem weapon trauma in two early medieval skeletal populations (AD 400-1200) from Ireland. *International Journal of Osteoarchaeology* 25 (3), 253–264.

Goodman, A.H., Brooke Thomas, R., Swedlund, A.C., Armelagos, G.J. 1988. Biocultural perspectives on stress in prehistoric, historical and contemporary population research. *Yearbook of Physical Anthropology* 31, 169–202.

Grauer, A.L. 1993. Patterns of anemia and infection from medieval York, England. *American Journal of Physical Anthropology* 91 (2), 203–213.

Grauer, A.L., Roberts, C.A. 1996. Paleoepidemiology, healing, and possible treatment of trauma in the medieval cemetery population of St. Helen-on-the-Walls, York, England. *American Journal of Physical Anthropology* 100 (4), 531–544.

Harman, M., Molleson, T.I., Price, J.L. 1981. Burials, bodies and beheadings in Romano-British and Anglo-Saxon cemeteries. *Bulletin of the British Museum of Natural History (Geology)* 35, 145–188.

Hemelrijk, E. 2004. *Matrona Docta: Educated Women in the Roman Elite from Cornelia to Julia Domna*. Routledge, London.

Holst, M. 2008. *Osteological Analysis, Heslington East, York*. Unpublished report no. 1108. Prepared for the Department of Archaeology at the University of York.

Holst, M. 2009. *Osteological Analysis, Heslington East, York*. Unpublished report no. 0609. Prepared for the Department of Archaeology at the University of York.

Holst, M. 2010. *Osteological Analysis, Heslington East, York*. Unpublished report no. 1010. Prepared for the Department of Archaeology at the University of York.

Holst, M. 2011. *Osteological Analysis, Yorkshire Museum, York*. Unpublished report no. 0610. Prepared for the Yorkshire Museum.

Hunt, D.R., Bullen, L. 2007. The frequency of os acromiale in the Robert J. Terry Collection. *International Journal of Osteoarchaeology* 17 (3), 309–317.

Hunter-Mann, K. 2005. *6 Driffield Terrace, York. An Assessment Report on an Archaeological Excavation*. Unpublished report no. 2005/55 for York Archaeological Trust.

Hunter-Mann, K. 2006a. More Romans lose their heads in York. Yorks. *Archaeology Today* 10, 1–2.

Hunter-Mann, K. 2006b. *Romans lose their heads: An unusual cemetery at The Mount, York*. Available at: http://www.iadb. co.uk/driffield6/driffield6.php (accessed 26th October, 2011).

Jacobi, K.P., Danforth, M.E. 2002. Analysis of interobserver scoring patterns in porotic hyperostosis and cribra orbitalia. *International Journal of Osteoarchaeology* 12 (4), 248–258.

Jones, R.F.J. 1984. The cemeteries of Roman York, in: Addyman, P.V., Black, V.E. (Eds.), *Archaeological Papers from York Presented to M. W. Barley*. York Archaeological Trust, York, pp. 34–42.

Judd, M. 2004. Trauma in the city of Kerma: Ancient vs. modern injury patterns. *International Journal of Osteoarchaeology* 14 (1), 34–51.

Lee, G.H., Cho, B.K., Park, W.J. 2010. A four year retrospective study of facial fractures on Jeju, Korea. *Journal of Cranio-Facial-Maxillary Surgery* 38 (3), 192–196.

Lewis, M.E. 2002. Impact of industrialization: Comparative study of child health in four sites from medieval and post-medieval England (A.D. 850–1859). *American Journal of Physical Anthropology* 119 (3), 211–223.

Lewis, M.E., Roberts, C.A., Manchester, K. 1995. Comparative study of the prevalence of maxillary sinusitis in later Medieval urban and rural populations in Northern England. *American Journal of Physical Anthropology* 98 (4), 497–506.

Lister, M. 1682. *Roman urns and other antiquities near York*. Royal Society Philosophical Collections IV, 87.

Lovell, N.C. 1997. Trauma analysis in palaeopathology. *Yearbook of Physical Anthropology* 40 (S25), 139–170.

Lovell, N.C. 2008. Analysis and interpretation of skeletal trauma, in: Katzenberg, M.A., Saunders, S.R. (Eds.), *Biological Anthropology of the Human Skeleton*. John Wiley and Sons, New Jersey, pp. 341–386.

Manchester, K. 1995. The palaeopathology of urban infections, in: Bassett, S. (Ed.), *Death in Towns. Urban Responses to the Dying and the Dead, 100–1600.* Leicester University Press, London, pp. 8–14.

Matthews, C.L. 1981. A Romano-British inhumation cemetery at Dunstable. *Bedfordshire Archaeology Journal* 15, 1–74.

Mattingly, D. 2007. *An Imperial Possession; Britain in the Roman Empire.* Penguin Group, London.

McIntyre, I., Holst, M. 2006. *89 The Mount, York; Assessment of skeletal remains.* OSA Report No. OSA05EX01. Unpublished report.

McIntyre, L. 2007. *Watching brief at Clifton Grange Hotel, York; Osteological assessment of skeletal remains.* OSA Report No. OSA07WB01. Unpublished report.

McIntyre, L. 2013. *Demography, diet and health in Roman York.* Unpublished PhD Thesis, University of Sheffield.

McKinley, J.I. 1992. *St. Stephens.* Unpublished report for Verulamium Museum.

McWhirr, A., Viner, L., Wells, C. 1982. *Romano-British Cemeteries at Cirencester.* Cirencester Excavation Committee, Cirencester.

Miles, A.E.W. 1994. Non-union of the epiphysis of the acromion in the skeletal remains of a Scottish population c. 1600 AD. *International Journal of Osteoarchaeology* 6 (2), 259–288.

Montgomery, J., Knüsel, C.J., Tucker, K. 2011. Identifying the origins of decapitated male skeletons from 3 Driffield Terrace, York, through isotope analysis: Reflections on the cosmopolitan nature of Roman York in the time of Caracalla, in: Bogonofsky, M. (Ed.), *The Bioarchaeology of the Human Head: Decapitation, Deformation, and Decoration.* University Press of Florida, Gainesville, pp. 141–178.

Müldner, G., Chenery, C., Eckardt, H. 2011. The 'Headless Romans': Multi-isotope investigations of an unusual burial ground from Roman Britain. *Journal of Archaeological Science* 38 (2), 280–290.

Museum of London Archaeology. 2009a. *Osteological Data for the Roman Southern Cemetery.* Available at: http://www.museumoflondon.org.uk/Collections-Research/LAARC/Centre-for-Human Bioarchaeology/Database/Roman+cemeteries/RomanSouth.htm (accessed 11th September, 2011).

Museum of London Archaeology. 2009b. *Osteological Data for the Roman Western Cemetery.* Available at: http://www.museumoflondon.org.uk/Collections-Research/LAARC/Centre-for-Human-Bioarchaeology/Database/Roman+cemeteries/RomanWest.htm (accessed 11th September, 2011).

Ortner, D. 1998. Male-female immune reactivity and its implications for interpreting evidence in human skeletal palaeopathology, in: Grauer, A., Stuart-Macadam P. (Eds.), *Sex and Gender in Palaeopathological Perspective.* Cambridge University Press, Cambridge, pp. 79–92.

Ottaway, P. 1999. York: The study of a late Roman colonia, in: Hurst, H. (Ed.), *The Coloniae of Roman Britain: New Studies and a Review.* Papers of the Conference held at Gloucester on 5-6 July, 1997. Journal of Roman Archaeology Supplementary Series No. 36, pp. 136–150.

Ottaway, P. 2004. *Roman York.* Tempus Publishing, London.

Ottaway, P. 2005. *1-3 Driffield Terrace, York. Assessment report on an archaeological excavation.* Unpublished report no. 2005/27 for York Archaeological Trust.

Pagnani, M.J., Mathis, C.E., Solman, C.G. 2006. Painful os acromiale (or unfused acromial apophysis) in athletes. *Journal of Shoulder and Elbow Surgery* 15 (4), 432–435.

Paley, K.J., Jobe, F.W., Pink, M.M., Kvintc, R.S., El Attrachec, N.S. 2000. Arthroscopic findings in the overhand throwing athlete: Evidence for posterior internal impingement of the rotator cuff. *The Journal of Arthroscopic and Related Surgery* 16 (1), 35–40.

Peck, J.J. 2009. *The biological impact of culture contact; a bioarchaeological study of Roman colonialism in Britain.* Unpublished PhD Thesis, Ohio State University.

Pfirrmann, C.W.A., Resnick, D. 2001. Schmorl's nodes of the thoracic and lumbar spine: Radiographic-pathologic study of prevalence, characterization, and correlation with degenerative changes of 1,650 spinal levels in 100 cadavers. *Radiology* 219 (2), 368–374.

Philpott, R. 1991. *Burial practices in Roman Britain. A survey of grave treatment and furnishing A. D. 43-410.* British Archaeological Reports, British Series 219. Tempus Reparatum, Oxford.

Pitts, M., Griffin, R. 2012. Exploring health and social well-being in late Roman Britain: An intercemetery approach. *American Journal of Archaeology* 116 (2), 253–276.

Redfern, R., Roberts, C. 2005. Health in Romano-British urban communities: Reflections from the cemeteries, in: *Fertile Ground: Papers in Honour of Susan Limbrey.* Symposia of the Association for Environmental Archaeology (22). Oxbow Books, Oxford, pp. 115–129.

Revell, L. 2005. The Roman life course: a view from the inscriptions. *Journal of European Archaeology* 8 (1), 43-63.

Roberts, C. 2000. Trauma in Biocultural perspective: Past, Present and Future Work in Britain, in: Cox, M., Mays, S. (Eds.), *Human Osteology in Archaeology and Forensic Science.* Greenwich Medical Media, London, pp. 325–356.

Roberts, C., Cox, M. 2003. *Health and Disease in Britain. From Prehistory to the Present Day.* Sutton Publishing, Stroud.

Royal Commission on Historical Monuments (England). 1962. *An Inventory of the Historical Monuments in the City of York.* Volume 1, Eburacum: Roman York. HMSO, London.

Schaefer, M., Scheuer, L., Black, S. 2009. *Juvenile Osteology: A Laboratory and Field Manual.* Academic Press, London.

Scheidel, W. 1995. The most silent women of Greece and Rome: Rural labour and women's life in the Ancient World. *Greece and Rome* 42 (2), 202–217.

Shepherd, J.P., Gayford, J.J., Leslie, I.J., Scully, C. 1988. Female victims of assault; a study of hospital attenders. *Journal of Cranio-Maxillofacial Surgery* 16, 233–237.

Shepherd, J.P., Shapland, M., Pearce, N.X., Scully, C. 1990. Pattern, severity and aetiology of injuries in victims of assault. *Journal of the Royal Society of Medicine* 83 (2), 75–78.

Simmonds, A., Márquez-Grant, N., Loe, L. 2008. *Life and Death in a Roman City – Excavation of a Roman Cemetery with a Mass Grave at 120–122 London Road, Gloucester.* Oxford Archaeology Monograph No. 6, Oxford.

Steckel, R.H., Larsen, C.S., Sciulli, P.W., Walker, P.L. 2006. *The Global History of Health Project Data Collection Codebook.* Unpublished manuscript. Available at: http://global.sbs.ohio-state.edu/new_docs/Codebook-01-24-11-em.pdf (accessed 21st July, 2011).

Stephenson, L.S., Holland, C.V., Cooper, E.S. 2000. The public health significance of Trichuris trichiura. *Parasitology* 121 (7), S73–S95.

Stirland, A. 2005. The Crew of the Mary Rose, in: Gardiner, J., Allen, M.J. (Eds.), *Before the Mast – Life and Death Aboard the Mary Rose.* The Mary Rose Trust, Portsmouth, pp. 516–562.

Sullivan, A. 2005. Prevalence and aetiology of acquired anaemia in Medieval York, England. *American Journal of Physical Anthropology* 128 (2), 252–272.

Sulosky, C. 2006. *A report on Romano-British cremated remains from York*. Unpublished report for the Yorkshire Museum.

Tucker, K. 2006. *Analysis of the inhumations, cremations and disarticulated human bone from sites at 1-3 Driffield Terrace, 6 Driffield Terrace and 129 The Mount*. Unpublished Report for York Archaeological Trust.

Walker, P.L. 2001. A bioarchaeological perspective on the history of violence. *Annual Review of Anthropology* 30, 573–596.

Warwick, R. 1968. The skeletal remains, in: Wenham L.P. (Ed.), *The Romano-British Cemetery at Trentholme Drive, York*. Ministry of Public Building and Works Archaeological Report No. 5, HMSO, London, pp. 112–165.

Watts, R. 2011. Non-specific indicators of stress and their association with age at death in Medieval York: Using stature and vertebral neural canal size to examine the effects of stress occurring during different periods of development. *International Journal of Osteoarchaeology* 21 (5), 568–576.

Webb, S. 2009. *Palaeopathology of Aboriginal Australians: Health and Disease across a Hunter-Gatherer Continent*. Cambridge University Press, Cambridge.

Wenham, L.P. 1968. *The Romano-British cemetery at Trentholme Drive, York*. Ministry of Public Building and Works Archaeological Report No. 5, HMSO, London.

Wheeler, H. 1985. The racecourse cemetery, in: Dool, J., and Wheeler, H., Roman Derby: Excavations 1968-1983. *The Derbyshire Archaeological Journal* 105, 222–280.

Wilson, A., Rackham, D.J. 1976. Parasite Eggs, in: Buckland, P.C. (Ed.), *The Environmental Evidence from the Church Street Roman Sewer System. The Archaeology of York. The Past Environment of York 14/1*. Council for British Archaeology, London, pp. 32–33.

Wood, J.W., Milner, G.R., Harpending, H.C., Weiss, K.M. 1992. The osteological paradox. Problems of inferring prehistoric health from skeletal samples. *Current Anthropology* 33 (4), 343–370.

York Archaeological Trust, 1973. *Bishophill II, 37, Bishophill Senior, York. Site 1973.15*. Archive notes. Primary site records kept by York Archaeological Trust.

York Archaeological Trust. 1981. *5, Rougier Street, York. Site 1981.12*. Archive notes. Primary site records kept by York Archaeological Trust.

York Archaeological Trust. 1984. *General Accident: 22–23 Tanner Row, York. Site 1983-4.32*. Archive notes. Primary site records kept by York Archaeological Trust.

York Archaeological Trust. 1987. *Leemans Garage, Wellington Row, York. Site 1987.24*. Archive notes. Primary site records kept by York Archaeological Trust.

York Archaeological Trust. 2011. *Gladiators: A Cemetery of Secrets*. Available at: http://www.yorkarchaeology.co.uk/headless-romans/index.htm (accessed 31st July, 2011).

Young, E.J. 1995. An overview of human brucellosis. *Clinical Infectious Diseases* 21 (2), 283–289.

The Riddle of the Sands: Sieving Methodology in the Recovery of Human Remains

Don P. O'Meara

This paper presents the results of a large scale sieving program that was undertaken as part of the excavation of a 19th-century crypt in Villiers Street, Sunderland, in northeast England. The paper discusses the role of sampling strategies within the collection of osteoarchaeological remains and how they compare with other fields, namely archaeobotany and archaeozoology. The results from the sieving program at Villiers Street are then presented, particularly the discussion of the numbers and types of bones most often collected in the sieve. It is argued that in the excavation and recovery of human bone a detailed knowledge of the human skeleton is essential to good excavation practice. It is also argued that further contributions of this nature from researchers in other geographical regions, and those working on material from other archaeological periods would bolster the osteoarchaeological community by providing empirical evidence for conducting sieving programs as an essential part of the collection of human remains, and not as an added luxury, or as something restricted to the those working in a 'research' environment.

Keywords Sieve; Villiers Street; Crypt

1. Introduction

'Tests indicate that excavators often miss far more than they realize and that the samples they recover are often so heavily biased as to be potentially misleading' (Payne, 1975).

Though Payne was discussing the recovery of large mammal bones his study indicates two issues to be highlighted in this paper. The first point is that caution must be taken when making inferences about sampling strategies when the effectiveness of this strategy is not based on empirical evidence. The second point is that due to differences in the way the disciplines developed there is arguably a much greater literature from the fields of archaeobotany and archaeozoology on sampling strategies based on empirical evidence than there is currently available from the field of osteoarchaeology; see for example the discussion in Barker and Worley (2014: 12) for details on archaeozoology sampling strategies and mesh size or general recommendations in English Heritage Environmental Archaeology Guidelines (Campbell *et al.*, 2011). The following case study presents the results of a sieving program that was undertaken as part of the excavation of a mid-19th century crypt in Sunderland, northeast England. It is hoped that the results of this particular case study will be of interest to those working in the excavation/exhumation of human remains in other geographical regions, and on material from different time periods. It is recognised that excavation in different geographical regions and excavations which deal with other cultural periods will have their own set of site formation processes. In this respect excavating crypts or graveyard sites is comparable to approaching the study of monastic sites, where they may have a shared origin and development, but are also subject to individual site natural and cultural formation processes (O'Meara, 2013: 290–93). It is hoped that this case study, due to the nature of the historical development of the site and the substrate from which the skeletons were excavated, will demonstrate the importance of sieving methodology to those legislating for, or directing, excavations and who are not themselves specialists in the field of osteoarchaeology.

To a certain extent informing osteoarchaeologists about the importance of sieving during excavation carries a note of preaching to the converted. However, osteoarchaeologists can learn from other fields in environmental archaeology (particularly archaeozoology and archaeobotany) by the manner in which these other disciplines use case studies on sampling and recovery to strengthen their arguments for systematic and thorough recovery strategies. To an extent these differing approaches stem from the nature of the archaeological evidence which means archaeozoologists and archaeobotanists have had to prove, with examples, why they want to sample certain volumes of sediment in particular frequencies. This is likely to be linked to the different scales at which other fields of environmental archaeology operate. Samples taken for environmental remains consider scales of importance on the context, site, local, regional and supraregional scale (O'Connor and Evans, 2005: 110–131). This has perhaps lead to a culture where environmental archaeologists need to justify to themselves that their strategies are effective, while also justifying to those funding the work (either from a commercial or research/academic source) that certain detailed strategies need to be employed in some circumstances. This concern with burial taphonomy as well as recovery has led to a healthy scepticism where 'One wonders whether there is any similarity between a published bone report and the animals exploited by ancient humans' (Davies, 1987: 23). In the case of human remains the presumption is that the sampling strategy is 100%, which is to say the aim on excavation is to recover all of the bones of the particular individual buried in a burial deposit. This can be seen in the thorough recommendations within current best practice for the excavation of human remains where advice is given to ensure 'no small bones or teeth have been overlooked in excavation' (McKinley and Roberts, 1999: 5). Though most excavation supervisors would accept the need for 100% recovery of human remains (at least we might hope so!), time and financial pressures may lead some to believe that as the skeleton will be arranged in a known anatomical position, in the case of inhumations, then the eye of the experienced excavator should be careful enough for full skeletal recovery. It is suggested here that this will only be acceptable with certain caveats. The discussion here is not the recommendation for a one way process aimed at highlighting issues in excavation methods and recovery methods. It must be remembered by all environmental archaeologists that in many cases our samples are taken by people working in difficult conditions and who are not specialists in our fields. As advanced as our techniques become, we need to feed back to the excavation process with continuous development for excavation staff, considerations of field techniques and how the field of osteoarchaeology operates when faced with commercial or time pressures. The roles of studies such as the one presented here are to show that a demonstrable knowledge of the human skeleton combined with an appropriate sampling strategy are needed to ensure the thorough recovery of human remains.

2. Case Study Background

The subject of this case study is a crypt excavation located under the Bethel Chapel, Villiers Street, Sunderland, in northeast England (Town, 2009; Storm and Buckberry, n.d.). As part of a phase of urban regeneration the site of the Bethel Chapel Sunday School (built in 1849 as an extension to the chapel constructed in 1817) was earmarked for demolition. This required the excavation of the inhumations which lay in the crypt beneath the location of the original Bethel Chapel. The excavation was undertaken as a joint project between North Pennines Archaeology (now Wardell-Armstrong Archaeology) and the Department of Archaeological and Environmental Science, University of Bradford.

The Bethel Chapel was built in 1817 to serve a population of Independent Congregationalists, whose origins lay in a split with the Presbyterian chapel on Robinson's Lane, Sunderland. This occurred during a period of rapid industrial growth when the populations of towns such as Sunderland were being swelled by immigrants from other parts of Great Britain (Mingay, 1986: 4–5). In the case of Villiers Street the arrival of Scottish immigrants led to an increase in the number of Presbyterian communities in the region; known as the 'Scottish Churches'. The early 19th century in Britain was a period of extensive chapel building fuelled by 'Old Dissent' Presbyterians, as well as the rapidly growing 'New Dissent' Methodists (Ryder, 2007). In terms of the religious culture in northeast England, Peter Ryder has highlighted that from the period of the mid-17th century through the 19th century Sunderland was a major centre of Nonconformist Christianity, though with a heritage asset that has been perhaps poorly studied until relatively recently (Ryder, 2012). As a Nonconformist chapel, the Bethel Chapel at Villiers Street is somewhat unusual in having a brick-built crypt system for burials. In this respect Ryder suggests comparable examples can be found at the Methodist chapels of the Bethesda Chapel at Hanley, Stoke-on-Trent and the Mint Methodist Chapel in Exeter.

After its construction in 1817, the chapel was enlarged in 1826 and it is this expansion that included the construction of the crypt. Over the course of the next 28 years 409 burials are recorded on the burial register. Due to the enactment of the 1852 Burials Act further registered inhumations ceased by 1854; the later history of the site as well as the main surviving features of archaeological interest, and the heritage asset of Nonconformist buildings in Sunderland have been studied by Buildings Archaeologist Ryder (Ryder, 2007; 2012). The enactment of the 1852 Burials Act is significant, as in order to comply with this legislation the crypt was filled with a layer of 'sanitising sand', in this case likely to be beach sand from the nearby coast (concluded from the occasional mussel and clam shells recovered during the sieving). This final layer of sanitising sand lay above the earlier sand layers which seem to have been built up to cover each new coffin internment, a sequence of activity interpreted from the excavated position of adult and juvenile coffins within the individual vaults (Town, pers. comms). It is this sand which would form the burial matrix from which the skeletons would be excavated, and from which the bones would be sieved. This arrangement can be commonly encountered in mid-19th-century crypts (Cox, 2002).

During the course of the excavation from January to April 2010 my role as on site environmental archaeologist involved the sieving of the burial sediment that was collected by the excavation team during excavation. From a sampling and osteoarchaeological point of view the site is interesting as it had a short period of burial activity (28 years; 1826–54), with a uniform burial pattern (inhumations in wood plank coffins), in a burial sediment that is generally uniform across the site (the sanitising sand). Variations in geology, changing burial practice, or the passage of time are not at this stage seen as factors which may have created differential preservation of skeletal elements, particularly when comparing this site to the analogous study at Whitby Abbey (Mays *et al.*, 2012). Differential preservation of soft tissue and coffin wood was noted, though how much this affected individual bones may only be clear at the completion of the analysis of the remains currently ongoing at the University of Bradford under Dr Jo Buckberry and Rob Janaway.

3. Sieving and Sampling Strategies

Before this project began, and through subsequent research, I felt that there was often only a brief mention of sieving in standard osteoarchaeological excavation texts. Coming from the perspective of experience with archaeobotanical and archaeozoological material I had anticipated greater emphasis (backed up with empirical case studies) on sample volumes, ideal sieve sizes and the nature of material most likely to be missed by excavators. In most cases the works cited below contain the entire reference to sieving practice within the texts. Brothwell does not mention sieving in his discussion on the excavation of human remains (Brothwell, 1981: 2–3), while Chamberlain mentions 'After all the visible bones have been lifted it is essential that the underlying few centimetres of soil are sieved in order to recover any fragments and small bones and teeth that may have been moved out of their original position' (Chamberlain ,1994: 54). Ubelaker advises 'All soil removed during excavation of a pit should be sifted through a fine screen to recover small artifacts and fragments of bone. Particular attention should be given to the recognition of infant bones, some of which are so small they are not likely to be discovered without screening' (Ubelaker, 1999: 19). Similar advice is given in the English Heritage Guidelines (English Heritage and Church of England 2005), while Roberts offers some practical, detailed advice on recovery

methods and recommends 'following lifting of all the obvious bones, the layer of soil below the skeleton is collected and sieved (1–2 mm mesh) to ensure no small bones or teeth, or fragments, have been missed' (Roberts, 2009: 78–79). A longer discussion on the importance of sieving is given by McKinley and Roberts (1999: 5), in their IfA Technical Paper. Within their paper the work of Keily (1989) is quoted. Her work on the Royal Mint site was presented at the 1989 meeting of the British Palaeopathology Association and though not published at the time it seems to be a similar examination to the current case study being presented here. Thus, though it is generally accepted that sieving is good practice it is rarely highlighted empirically why it is superior to hand collection alone.

The identification of the effect of incomplete excavation of human skeletons was the focus of a study by Simon Mays for material relating to a skeletal assemblage from Ipswich, England (Mays, 1992). In this case Mays was studying a skeletal assemblage where no sieving had taken place but used the evidence from the bone preservation to try to assess whether the loss was due to possible excavation methodology or taphonomic factors associated with the burial conditions. At Whitby, Mays *et al.* (2012) were faced with poor preservation in a clay soil, with burials ranging from the 7th–9th century AD. In this case Mays and colleagues used the opportunity to test human remains recovery methods under these conditions and it is hoped that the case study being presented here can act as a comparable example with the experiences at Whitby, and can begin to fulfil some of the suggestions presented in that paper (Mays *et al.*, 2012: 3253).

4. Comparisons with other fields

Though a direct comparison between sampling for human remains and for other classes of ecofacts (such as animal bone and plant remains) cannot be made on a methodological basis (cf. Mays *et al.,* 2012: 3248), it is interesting how these fields treat the recovery of their research material. In the introduction to the English Heritage Guidelines on the treatment of human remains from Christian burial sites Simon Thurley reminds us that 'Human remains are a focus of religious beliefs and notions of decency and respect for the dead, as well as arousing great public interest' (Mays, 2005: 2). It is perhaps these beliefs that mean archaeologists, and the general public, accept that a 100% collection policy is necessary. Any archaeologist working in a professional capacity would (we hope) not willingly ignore human remains as a skeleton was being excavated. This does not, however, acknowledge that material can be missed if it is only subject to hand collection. In contrast, it can be suggested that it was the belief that animal bone and plant remains were being routinely poorly collected, or not sampled at all, which encouraged archaeozoologists and archaeobotanists to seek out case studies to reinforce their argument as to why sieving, or other systematic sampling, should be undertaken. This highlighted not just what was being missed but also the potential archaeological information that can be lost. An authoritative list of these studies is beyond the scope of this paper but some useful case studies will be outlined briefly for different classes of ecofacts.

In the case of archaeozoology the importance of sieving and the bias of hand recovery have been examined through systematic studies for several decades. To an extent this may be the product of the change towards New Archaeology as it was taken up by environmental archaeologists in the 1960s–70s; what was called, in relation to the Experimental Earthwork Project, the 'currents in contemporary archaeology … in theory and method, in organization and personnel, and in changes externally in the climate of research and internally in the growth of professionalism' (Bell *et al.*, 1996: xix). An example of this type of work includes Payne's examination of skeletal assemblages from Greece (Payne, 1975). Though talking about animal bones Payne touches on some of the topics relevant to the recovery of human remains when he challenges the notion that 'it has generally been implicitly accepted that excavators recover all or nearly all of the identifiable bones of the large mammals and thus that numerical data for the larger mammals can be used with confidence' (Payne, 1975: 7). In his study Payne uses empirical tests to demonstrate the biases of hand recovery and has provided archaeozoologists with a sampling model that can still be used today. This has also been developed and expanded by other researchers; for example the recovery of animal bones at 46–54 Fishergate, York (O'Connor, 1988). Thorough examples of recovery issues are discussed in standard texts and include the recovery of bird bones (O'Connor, 2000: 33–35; Serjeantson, 2009: 100–102) and fish bones (Wheeler and Jones, 1989; Zohar and Belmaker, 2005). It is an example of the perceived importance of recovery methodology that Wheeler and Jones devote 20 pages of their publication (or 10% of the whole book) to a section on 'Methods of Recovery' (1989: 38–60). In relation to bird bone, Serjeantson discusses not only the importance of sieving but quotes various papers which recommend various mesh sizes (Serjeantson, 2009: 101).

From an archaeobotanical perspective, attempts to provide a standard sampling tactic, encompassing sampling sizes and processing methodology, can be seen in Green's work in southern England (Green, 1979). By 1985 this had developed into a lively discussion on manual processing (Kenward *el al.*, 1980; Badham and Jones, 1985) and sample sizes for recovered remains (van der Veen and Fieller, 1982). The usefulness of these studies is that they provide case studies that other environmental archaeologists can use to implement their own sampling methodologies by showing evidence to their site director or legislating individual (who in the 1980s was unlikely to be an experienced environmental archaeologist). This discussion continues within archaeobotany and is aided (certainly from the perspective of the archaeobotanist or archaeozoologist),

by the publication of the English Heritage Environmental Archaeology Guidelines (Campbell *et al.*, 2011; Barker and Worley, 2014). Within these documents case studies present empirical data on the problems of biased recovered. This includes a discussion on the biases of sieved and non-sieved fish remains (Stallibrass, 2011), and the problems of partial sample assessment from archaeobotanical bulk samples (Huntley, 2011). An example of this bias was demonstrated from 25 Bridge Street, Chester where it was demonstrated that relying on hand collection not only led to the recovery of fewer fish bones, but led to a bias in the diversity of species recovered (Jacques *et al.* 2008). It is highlighted that when sieving was employed it led to a much greater number of bones to be recovered (90 hand recovered bones compared to 3638 sieved bones), and a greater diversity of species (11 species are represented in the hand collected bones, while 23 are represented in the sieved material) (Stallibrass, 2011: 30–31). From an archaeobotancial perspective Huntley discussed the importance of processing a full suite of samples from an excavation, rather than further sub-sampling during the post-excavation process. In this case study an initial assessment of 30 samples out of 110 was chosen. However, when sent for full analysis the results of the 30 samples bore little resemblance to the full processing of the 110 samples (Huntley, 2011: 29). Some experimental work has also been conducted for the field of archaeobotany in de Moulins' examination of the effect of different pre-treatment methods on the recovery of charred plant remains (de Moulins, 1996). Here de Moulins examined the effects of different pre-treatment methods, on various different sediment types, and using different flotation tank types. In these case studies the emphasis is not just that incomplete sampling leads to a reduction in the volume of material recovered, but also leads to a reduction in the diversity of the remains.

5.   Results of the Sieving

The assemblage examined by sieving at Villiers Street consisted of 198 burials, and over 14,000 litres of sanitising sand. This was dry sieved through a 3 mm wire mesh, which it was hoped was sufficiently small in aperture to retain the bones of the excavated skeletons (infant skeletons were block-lifted and sieved off-site), while being sufficiently large to allow for timely and efficient sieving. The results presented here concern the work of twelve excavators who were on the project team from between less than one month, to those who worked for the entire project length of approximately four months. Each excavated between five and forty-seven burials. Of interest here is not only how the assemblage as a whole was formed, but also how the background of each individual excavator might affect the recovery of different types of human remains. This analysis was possible as by using the site context records the identity of the individual who excavated each skeleton could be recorded with the bones sieved from that particular skeleton context. This was of interest as the twelve excavators had a variety of backgrounds within archaeology, both academically, and in terms of their field experience. The twelve excavators can be divided into three distinct groups:

1. Group A: those with an MSc in Human Osteoarchaeology, and 1–2 years field experience.
2. Group B: those without an MSc and limited continuous field experience (1–2 years)
3. Group C: those with over 10 years field experience, none of whom had an MSc.

In particular I was interested the difference between the academic qualification, particularly the MSc in Osteoarchaeology, and field experience of the group, and how it might affect their recovery of the bones. The results of this study are presented in Table 9.1. The data can be interpreted in multiple ways; and speaking to those at the BABAO Conference 2013 in York raised various other valid suggestions.

From the perspective of Body Part Analysis (BPA) I looked at the number of bones recovered and divided this by the number of times that bone occurs in the (adult) human skeleton, a methodology that might be more familiar to archaeozoological quantification than quantification in human osteoarchaeology (O'Connor, 2000: 68–79). This will average out the issues that might be present with some elements such as teeth, compared to individual carpals or tarsals. This was not calculated for the bones of the skull or the pelvis as based on the time I had to examine the material I could not determine whether I was examining multiple fragments of a single bone, or separate fragments of different bones. In Table 9.1 the individual excavators are labelled A-1 to C-3 based on the division of the twelve people previously presented. The difficulty in studies of this nature is to avoid the sense of judging a particular individual's abilities; therefore it would be unfair to print names or initials.

The total number of bones recovered from sieving (via fragments and bones identified to their anatomical position) was 2422. The number of identified bones and fragments recovered per individual excavator was an average of 201 bones but this ranged from 19 for one excavator to 1167 for another excavator; remembering of course that there was a range of excavated inhumations from 5–47 across the team.

From the perspective of the BPA it can be seen that the long bones fare quite well with the femur and the humerus being the least commonly recovered bones in the sieve, while the tibia, radius and ulna are found occasionally (it should be pointed out that these were recovered as identified fragments rather than whole bones). It can also be seen that the 93 teeth recovered are placed relatively low down on the scale of recovered bones (this is of course assuming a compliment of 32 teeth). At the other end of the scale it is noticeable that the phalanges are quite commonly missed, even

when being balanced by dividing the total number by 56 to represent both hand and foot phalanges (at the time as I was not fully happy with my ability to separate all foot from all hand phalanges therefore I recorded them together). By far the most commonly missed bone was the patella. This must be considered in relation to some post-burial disturbance (see below) but is a timely reminder of the ability of this bone to be missed. Looking at the carpals there was some variation in the frequency of each bone recovered from sieving, though why the hamate was so frequently recovered is somewhat unusual. Looking at the tarsals recovered it is perhaps more obvious why a distinctive bone like the calcaneous was spotted during excavation and not therefore available to be recovered by sieving. The frequency of retrieved cuneiforms was, I thought, quite high, though based on the BPA this places them in a similar position to the navicular.

When looking at these results based on the individual excavators it needs to be accepted that there was a variation in the number of excavated skeletons per person from 5–47 (some individuals had roles onsite other than excavation, such as surveying or conserving coffin plates). In order to account for the differences in the number of skeletons excavated, the total number of bones sieved from the contexts excavated by an individual was divided by the number of skeletons excavated by that individual. In a hypothetical ideal situation this would produce consistent zeros, as all the bones would be collected by the excavator. The higher the number the more bone was being missed on average per skeleton during excavation. What is not known at this time is what a 'normal' or average number might be, or even if it is possible to devise an 'acceptable' level of recovery. The results can be broken down as follows:

1. Group A: the three excavators with an MSc all scored less than 4.2
2. Group B: those with 1–2 years experience scored from 4–9.8
3. Group C: those with over 10 years field experience, and who have excavated numerous inhumations scored between 10.8–24.8

6. Interpreting the Results

Initially I proposed several possible reasons for this:

1. Group A: the MSc students, being more familiar with the skeletal anatomy are more conscious about what bones they should be expecting while they are excavating.
2. Group B: those without a formalised knowledge of skeletal anatomy and little experience are more cautious and take more care.
3. Group C: those with more field experience were more comfortable excavating human remains quickly, though frequently missed the smaller bones.

Don P. O'Meara

Table 9.1 *Summary of the material recovered through sieving*

	Totals	BPA	A-1	A-2	A-3	B-1	B-2	B-3	B-4	B-5	B-6	C-1	C-2	C-3
Number of skeletons			6	5	5	11	12	25	22	16	5	34	10	47
Numbers of bones			19	20	21	43	61	208	188	138	49	366	142	1176
Fragments	990		3		13	13	21	108	64	98	4	196	50	420
Identified bones			16	20	8	30	40	100	122	40	45	170	92	747
Bones/Contexts			3.2	4	4.2	4	5.1	8.3	8.5	8.6	9.8	10.8	14.2	24.8
ID Bones/Contexts			2.7	4	1.6	2.7	3.3	4	5.5	2.5	9	5	9.2	15.9
Skull	117				4		6	23	4		1	16	4	59
Mandible	4	4											2	2
Teeth	93	2.9	1		1	2	2	2	5	4	2	23	1	50
Hyoid	2	2												
Scapula	7	3.5		1					1					5
Clavicle	5	2.5		1									2	3
Vertebra	90	3.8	1	1				3	16	5	5	6	15	38
Sacrum	3	3												3
Coccyx	8	8								1				7
Sternum	3	3							1			1		1
Ribs	134	5.1	1	9		5	1	7	3	2	1	14	13	78
Pelvis	19					1			1			1		16
Humerus	2	1												2
Radius	5	2.5			1				1			1		2
Ulna	8	4								2				6
Hammate	22	11				1			1	1	1	1		17
Capitate	6	3							1		1			4
Lunate	13	6.5								2				11
Pisiform	10	5							2					8
Scaphoid	13	6.5						1		1	1	1	1	8
Triquetrum	7	3.5						2					1	4
Trapezium	15	7.5			1	1		1				1		11
Trapezoid	10	5					1	1						8
Metacarpal	76	7.6				1	2	2			1	9	6	55
Phalanges	568	10.1	13	6	1	14	24	37	70	19	28	76	27	253
Femur	3	1.5										1		2
Patella	38	19				2		1	3	1		1	3	27
Tibia	4	2							1					3
Fibula														
Calcaneus	3	1.5					1							2
Navicular	19	9.5				1		3	2		1	1	1	10
Talus	8	4				1		2	1				1	3
Cuboid	12	6						1	1			2	1	7
Cuniform	57	9.5		1			2	7	2	1		5	8	31
Metatarsal	24	2.4					1	2	5			6	6	4
Longbone	19						1	1	4	1	2	4		5
Cartilage	5			1								2		2

Key: *Numbers of skeletons*: the number of skeletons excavated by each individual (A-1 to C-3) for which there is a record for the sieved material. *Numbers of bones*: the total numbers of bones recovered by sieving for the skeletons excavated by this individual. *Fragments*: the numbers of unidentified fragments recovered from the sieved material for each individual. *Identified bones*: the numbers of identified bones recovered from the sieved sand from each individual. *Bones/Contexts*: the average number of total bones per context. *ID Bones/Contexts*: the average number of identified bones per context. *BPA*: Body Part Analysis.

However, a number of alternative explanations can be suggested. Those in Group C, knowing that there was an active sieving process, did not spend time looking for each individual phalanx as they assumed they would be picked up in the sieve. It should also be pointed out that from a logistical point of view this group excavated the most skeletons.

It was asked at the 2013 conference whether those with more experience were given more fragmentary and 'difficult' remains to deal with, however, when recalculated with or without the inclusion of the fragmentary remains the results are largely the same. In any case the excavation staff were given individual vaults to work on and therefore the allotment of skeletons was random. There are of course more taphonomic factors in play than the work of a particular individual. Indeed, one of the difficulties onsite was the process whereby three or four coffins which had been stacked on top of each other (between 1826–54), collapsed at some point after they were buried in the sand. The final layer of sanitising sand happened as a single archaeological 'event' in 1854, however the collapse of the coffins did not occur as a single event, or in an even nature. The coffins appeared to have collapsed first in the area of the thoracic cavity, presumably as fluids in the thoracic cavity leaked from the body. This created a sag in the skeleton as the wood around the midpoint of the coffin degraded first, leading to the thoracic bones being physically lower than the head or foot bones. The collapse had the effect of making the skeletons physically much closer, and occasionally impact upon each other to the extent that in certain cases when excavating the thoracic cavity of one skeleton elements of the skeleton stratigraphically below would be found within it. Again, it is important to visualise the coffins collapsing in an uneven way, with the mid-point of the coffin often collapsing first, rather than a series of coffins falling together with a 'concertina' effect. The collapse of the coffins may also have led to the movement of certain bones. The loss of the patella from its anatomical position may be a result of this. During the excavation I was shown multiple examples where the excavator had spotted a patella from an individual many centimetres from its original anatomical position. Though these factors created difficulties for all of the excavation staff, on balance it is felt that the single clearest explanation is that the knowledge of human anatomy possessed by those with an MSc was the main factor in their more complete recovery of remains. As these individuals excavated a skeleton they were more conscious about the numbers of bones they should be expecting.

Two suggestions to the issues discussed in this paper could include providing each excavator with a geological sieve or making them more aware of the individual carpals and tarsals through training. McKinley and Roberts (1999: 2) suggest practical and affordable activities such as providing a printed/poster skeleton chart in the site cabin with details of the human skeletal anatomy. One change which might be considered is to think about the labelling or presentation of carpals and tarsals on skeleton sheets. To people, such as those with an MSc in this study, each of the representations on the skeleton sheet represents a named bone which they are familiar with. For archaeologists who have not been given training on the individual named elements of the skeleton these bones might as well be represented as triangles or circles. If they were aware of the names of the individual bones and the features of these bones they would, I think, become more mindful of the presence or absence of each bone as they were excavating.

The taphonomic elephant skeleton in the room is of course my personal ability to identify the bones as they were appearing in the sieve; a skill likely to have been helped by my hours of examining heavy residues from flotation processing. My knowledge of human skeletal anatomy stems from an undergraduate training class, for which I was later a teaching assistant while completing my MA studies. My current interest in bones is in the field of archaeozoology rather than in human remains studies, though I would be confident in my ability to identify a human bone and determine whether it was a left or a right. During the sieving process I referred to my copy of White and Folkens (2005), and was able to call upon the abilities of the numerous onsite osteoarchaeologists for identifying material such as ossified cartilage. However, ultimately the material I was discarding from the sieve was not itself being double checked by another person, thus to answer the question posed by Juvenal: there was no-one to watch the watchman. Everyone in the excavation process from onsite personnel to post-excavation is a taphonomic agent and it is by engaging with this reality that errors or omissions can be highlighted.

7. Conclusion

We heard at the start of the 2013 conference from Martin Smith who commented on Pitt-Rivers' excavations, and how only the long bones and skull seemed to have been retained in the archive. Pitt-Rivers had no research questions based on the other elements and therefore he was probably content to not retain them. This early pattern of selective antiquarian bone recovery has been noted for a number of other sites and reflects the accepted research questions of the day (Smith and Brickley, 2009: 73). As our research questions develop we need to be aware of what is lost as much as we need to examine the remains being studied. If excavation directors are happy not to sieve, and this is approved by the curator or county/city archaeologist, then that is their prerogative. However, they should not be complacent thinking that they are operating within best practice. Nor should they blame the non-recovery of certain elements on that wonderfully vague term 'due to taphonomic reasons'. Also, those legislating for or protecting archaeological sites should be more proactive about assessing the sampling strategy for exhumations and incorporating it into the written scheme of investigations at the pre-excavation stage. There are a number of ethical and practical benefits to the value of sieving strategies, and in the sampling methodologies highlighted by certain texts (McKinley and Roberts, 1999; Mays *et al.*, 2012). From an ethical point of view it leads to a more complete recovery of human remains and avoids situations where bones might end up on the spoil heap. This should be seen as a major point of interest in regions of the world where the exhumation of human remains can often carry politically and socially loaded meanings to members of the public.

From a practical point of view there are two main benefits. Firstly, it ensures that a more complete rate of skeletal recovery is achieved. Though the main work of aging and sexing a skeleton will be achieved via the main bones of the limbs, pelvis and skull if this is the limit of the examination then the archaeological community might as well adopt the sampling methodology of Pitt-Rivers. The study of pathological skeletal elements such as degenerative joint diseases would be hampered if an incomplete collection of carpals, tarsals and phalanges takes place. It can also be argued that if the typical elements of the human skeleton are not recovered then elements such as kidney stones or extra sesamoids are also likely to be missed, with implications for the study of disease patterns in past human populations. Secondly, it is suggested that the excavation of human remains is an important touchstone for techniques of archaeological excavation in general. From a practical point of view if human remains cannot be thoroughly sampled then it raises questions about the sampling strategies for other classes of material. Though those studying other elements of the environmental record or the artefactual material might not be interested in the results of detailed analysis of the human remains, they should take note of the general preservation and recovery rates of the human skeletal material as a means of gaining an insight into the general sampling strategies employed onsite (while acknowledging of course that there will be recovery factors unique to their own particular classes of material). This might be particularly significant when re-evaluating older archived projects. In a project where skeletons have been excavated and where the full complement of bones is not present the non-osteoarchaeologists should consider where that places the recovery rates for their own material.

The IFA working paper by McKinley and Roberts (1999) deals with all of the issues raised in this paper, but lip service should not be paid to recommendations of this nature, or to documents such as the English Heritage Environmental Archaeology Guidelines (Campbell *et al.*, 2011). Finally, from the biological anthropology community (or indeed to archaeobotanists and archaeozoologists etc.) there must be an ongoing evangelisation to excavation staff to promote the field of human osteoarchaeology and why the methods of excavation of this material are important. As already mentioned above, but worth repeating; it must be remembered by all environmental archaeologists that in many cases our samples are taken by people working in difficult conditions and who are not specialists in our fields. In this context they must be regarded as an active part of the research process, with a stake in the planning and decision making process. Complacency in this field can often develop where individuals feel that the 'specialist' is not concerned with the onsite procedures and is only interested in their own particular field divorced from the context in which it is was found.

Acknowledgements

Thanks to Peter Ryder for his help on the historical background for this site, and to Matt Town, North Pennines Archaeology Project Manager for the Villiers Street Project for commenting on my conclusions in this paper. Also, to Megan Stoakley for answering my pestering questions about diseases, to Jocelyn Strickland for her onsite double-checking of my identifications and particularly to the team from Bradford University, namely Rob Janaway, Dr Jo Buckberry, Dr Andrew Wilson and Andy Holland. A final thanks to Dr Barra O'Donnabhain of University College Cork for introducing me to the field of osteoarchaeology.

References

Badham, K., Jones, G. 1985. An experiment in manual processing of soil samples for plant remains. *Circaea* 3 (1), 15–26.

Barker, P., Worley, F. 2014. *Animal Bones and Archaeology: Guidelines for Best Practice*. English Heritage, Swindon.

Bell, M., Fowler, P.J., Wilson, S.W. 1996. *The Experimental Earthwork Project, 1960–92*. CBA Research Report: Council for British Archaeology, York.

Brothwell, D.R. 1981. *Digging up Bones: The Excavation, Treatment and Study of Human Skeletal Remains*. Third Edition. Cornell University Press, New York.

Campbell, G., Moffet, L., Straker, V. 2011. *Environmental Archaeology: A Guide to the Theory and Practice of Methods, from Sampling and Recovery to Post-excavation*. 2nd Edition. English Heritage, Swindon.

Chamberlain, A. 1994. *Human Remains*. British Museum Press, London.

Cox, M. 2002. *Crypt Archaeology: An Approach*. IFA Technical Paper No.3. Institute of Field Archaeologists, Reading.

Davis, S.J. 1987. *The Archaeology of Animals*. Routledge, London. 2005 Digital Reprint.

English Heritage and Church of England. 2005. *Guidance for Best Practice for Treatment of Human Remains Excavated from Christian Burial Grounds in England*. English Heritage, London.

Green, F.J. 1979. Collection and interpretation of botanical information from medieval urban excavations in Southern England, in: Korber-Grohne, U. *Festschrift Maria Hopf*. Archaeo-Physika 8. Rheinland Verlag, Köln, pp. 39–55.

Keily, F. 1989. *On-site recovery of human bones*. Paper presented at the one day meeting of the Palaeopathology Society, May.

Kenward, H. K., Hall, A. R., Jones, A. K. G. 1980. A tested set of techniques for the extraction of plant and animal macrofossils from waterlogged archaeological deposits. *Science and Archaeology* 22, 3–15.

Mays, S. 1992. Taphonomic factors in a human skeletal assemblage. *Circaea* 9, 54–58.

Mays, S., Vincent, S., Campbell. G. 2012. The value of sieving of grave soil in the recovery of human remains: an experimental study of poorly preserved archaeological inhumations. *Journal of Archaeological Science* 39, 3248–3254.

McKinley, J.I., Roberts, C.A. 1999. *Excavation and Post-Excavation Treatment of Human Remains*. Technical Paper Number 13. Birmingham, Institute of Archaeologists.

Mingay, G.E. 1986. *The Transformation of Britain 1830–1939*. Routledge & Kegan Paul, London.

Moulins, D. de. 1996. Sieving experiment: The controlled recovery of charred plant remains from modern and archaeological samples. *Vegetation History and Archaeobotany* 5, 153–156.

O'Connor, T.P. 1988. *Archaeological Bone Samples Recovered by Sieving: 46–54 Fishergate, York, as a case study*. Ancient Monuments Laboratory Report 190/88. English Heritage, London.

O'Connor, T.P. 2000. *Animal Bones in Archaeology*. Sutton Publishing, Stroud.

O'Connor, T.P., Evans, J.G. 2005. *Environmental Archaeology: Principles and Methods*. Sutton Publishing, Gloucestershire.

O'Meara, D. 2013. Scant Evidence of Great Surplus: Research at the rural Cistercian Monastery of Holme Cultram, Northwest England, in: Groot, M., Lentjes, D., Zeiler, J. (Eds.), *Barely Surviving or More than Enough?* Sidestone Press, Leiden, pp. 279–296.

Payne, S. 1975. Partial recovery and sample bias, in: Clason, A.T. (Ed.), *Archaeozoological Studies*. Elsevier, New York, pp. 7–17.

Roberts, C.A. 2009. *Human Remains in Archaeology: A Handbook*. CBA Practical Handbook 19. Council for British Archaeology, York.

Ryder, P.F. 2007. *Bethel Chapel, Villiers Street, Sunderland*. Unpublished Building Recording.

Ryder, P.F. 2012. *Nonconformist chapels and meeting houses in Sunderland*. Unpublished Report for Tyne and Wear Council.

Serjeantson, D. 2009. *Birds*. Cambridge University Press, Cambridge.

Smith, M., Brickley, M. 2009. *People of the Long Barrows: Life, Death and Burial in the Earlier Neolithic*. The History Press, Gloucester.

Storm, R., Buckberry, J. (n.d.). *Assessment Report and Proposal for Post-Excavation Analysis of the Human Skeletal Remains from the Bethel Chapel Crypt, Villiers Street, Sunderland (VSS-A)*. Report prepared for Homes and Communities Agency. Biological Anthropology Research Centre, University of Bradford.

Town, M. 2009. *Bethel Chapel Crypt, 12–14 Villiers Street, Sunderland*. Unpublished project design for archaeological excavation. North Pennines Archaeology VSS-A CP991/09.

Ubelaker, D.H. 1999. *Human Skeletal Remains: Excavation, Analysis, Interpretation*. 3rd Edition. Taraxacum, Washington, D.C.

van der Veen, M., Fieller, N. 1982. Sampling seeds. *Journal of Archaeological Science* 9, 287–298.

Wheeler, A., Jones, A.K.G. 1989. *Fishes*. Cambridge University Press, Cambridge.

White T.D., Folkens, P.A. 2000. *The Human Bone Manual*. Elsevier Academic Press, Massachusetts.

Zohar, I., Belmaker, M. 2005. Size does matter: Methodologcial comments on sieve size and species richness in fishbone assemblages. *Journal of Archaeological Science* 32, 635–641.

Biological Age Estimation of Non-adult Human Skeletal Remains: Comparison of Dental Development with the Humerus, Femur and Pars Basilaris

Carla L. Burrell, Carole A.L. Davenport, Raymond J. Carpenter and James C. Ohman

Estimates for age at death in non-adult remains are based on ontological changes reflecting physiological age. Physiological age can be calibrated with chronological age by analyses of reference populations with known age at death. In turn, the calibration for chronological age is birth, but this event can occur at different physiological ages and is one source of the variability displayed at any given chronological age. To investigate the variability present within an unknown population and produce population specific growth standards, the dental calcification and eruption rates were compared with the developmental rates of the humerus, femur, and pars basilaris.

In this study, a sample of 171 medieval non-adult skeletons was drawn from the Poulton Research Project. Their dental radiographs were seriated into developmental order to establish relative physiological age. Seriated dental developmental age was compared with age based on measurements obtained from each humerus, femur and pars basilaris.

For all metrics, Kendall's T and Spearman's ρ correlations were high and significant (p < 0.001). Growth appears to be curvilinear in all cases, with accelerated growth during early childhood. For humeri and femora, there is a second, later period of accelerated diaphyseal growth that coincides with the eruption of the permanent second molar. This study demonstrates that using dental seriation as a proxy for developmental age is a promising approach for examining differential growth and likely chronological age in archaeology and forensic anthropology.

Keywords Developmental Age; Chronological Age; Seriation

1. Introduction

An accurate assessment of an individual's age at death is an important life history parameter in both archaeology and forensic anthropology. When working with non-adult archaeological skeletal remains, the most reliable methods for age at death estimations are derived from dental development studies (Ubelaker, 1999; Cardoso, 2006; AlQahtani *et al.*, 2010). However, a common issue when working with archaeological material is the preservation and completeness of non-adult skeletal remains. This provides a means for researchers to evaluate the remaining crania and postcranial material in order to estimate age at death. Some consider the use of epiphyseal union and fusion of skeletal elements (Scheuer and Black, 2000) to assess age at death. However, due to the fragility of such epiphyseal elements they may be damaged or absent due to various extrinsic factors (e.g., excavation

methods and taphonomic processes), which prevents the application of these methods (Lewis, 2007). Age at death is often estimated from long bone lengths due to the simple methodology and frequent use of this material in numerous studies both from modern data with known age (Maresh, 1970; Hoffman, 1979) and archaeological data with estimated ages (Rissech *et al.*, 2008; Primeau *et al.*, 2012). However, there is further discrepancy between the estimation of age through the application of data to a different population and/or time period.

For the non-adult skeleton, ontogenetic changes are relatively similar between individuals, although epigenetic factors such as health and nutrition can introduce some variability (Lampl *et al.*, 1992; Lampl and Johnston, 1996; Cameron and Demerath, 2002; Langley-Evans, 2015). However, age at death analyses of non-adult skeletal remains are more accurate than estimations for adults because the former are derived from these relatively consistent ontological changes. Estimates for age at death of non-adult remains are always derived from developmental, or physiological, ages (i.e., the length of time since fertilisation). Developmental ages can be transformed into chronological ages (i.e., the length of time since birth) by comparisons with reference populations with known chronological ages at death. This transformation is done because chronological age is considered the standard method to establish the age of living individuals. However, chronological age assessment is based more in culture than it is in biology (Belsky *et al.*, 2015). Birth is actually a somewhat 'arbitrary zero calibration point' for age along the developmental timescale. In contrast, the 'zero calibration point' for developmental age is 'absolute' because it always occurs at fertilisation.

In a recent sample that controlled for 'normal' healthy pregnancies, Jukic *et al.* (2013) reported the mean time from ovulation to birth was 268 (± 3.4) days, with a range from 247 to 284 days. This 37-day range for live singleton births occurred despite exclusion from the sample of pre- and post-term births, as well as those with medical intervention. Nevertheless, this likely typical 37-day range still accounted for 13.8% of average gestation length. Jukic *et al.* (2013) also reported that the time from ovulation to implantation varied from six to twelve days, with 42% occurring on day nine, and showed that this variability was predictive of total gestation length. This suggests that there is also some variability in developmental rates between individuals.

Jukic did not address developmental age at birth. However, given the marked variability in gestation length, it is likely that there is variability in developmental age at birth; that is, birth can occur at differing developmental ages. Therefore, birth is an 'arbitrary zero calibration point', and this accounts for much of the variability observed in developmental age at any given chronological age. In other words, the variability in the 'arbitrary zero calibration point' of birth creates pseudo-variability in developmental age at any given chronological age. Therefore, developmental or physiological age can provide a more fundamental depiction of ontological changes and permits a more direct comparison of relative growth differences between skeletal elements.

Dental development is one indicator of skeletal maturity that is well documented. For any individual, development of the dentition is a continuous process that begins during embryonic life and continues until late adolescence (Schour and Massler, 1941; Moorrees *et al.*, 1963a, b; Demirjian *et al.*, 1973; Ubelaker, 1987, 1999; AlQahtani *et al.*, 2010). This process progresses through two usually overlapping stages of development; that is, the deciduous dentition followed by the permanent dentition.

Lewis and Garn (1960) argued that calcification rates in teeth are only minimally affected by environmental conditions (i.e., epigenetic factors). Anderson *et al.* (1964) agreed, suggesting that dental development stages

are less affected by variation in nutrition and endocrine status. However, Cameron and Demerath (2002) suggested that epigenetic factors could introduce some variability during 'critical periods' of growth. Nevertheless, the pattern of dental development is recognisable across all populations and proceeds at a reasonable and predictable rate in both modern and archaeological contexts (Stull *et al.*, 2014). Therefore, dental development is one of the most accepted methods for assessing developmental age, as well as its transformation into a chronological age estimate. Sometimes, however, the non-adult dentition is incomplete or missing, but other skeletal elements are preserved and can be used to assess developmental age.

The *pars basilaris* is one of the four main centres of ossification for the occipital. It is the most robust of the four and often preserved in fragmentary human remains. The *pars basilaris* undergoes endochondral ossification and becomes recognisable during the foetal period (Scheuer and MacLaughlin-Black, 1994). In addition to absolute growth, its shape changes due to relative growth differences (Scheuer and Black, 2000). Thus, various metrics obtained from the *pars basilaris* can be used to assess developmental age.

Long bones also form by endochondral ossification, with the diaphysis developing from the primary centre of ossification. Therefore, diaphyseal lengths can be used to assess developmental age. Relative to the remaining postcranial skeleton, the humeral and femoral diaphyses are commonly recovered from archaeological contexts due to their robust nature (Willey *et al.*, 1997). Although measurements were taken for all long bones, at present the study focuses on those that were most often retrieved during the excavation.

This research explores *pars basilaris* growth and humeral and femoral diaphyseal lengths relative to dental development for a sample of 171 non-adult remains recovered from a Medieval church graveyard at Poulton, Cheshire. Osteological research on this large human skeletal collection is the first to take place for this sample. The aim of this study is to seriate the radiographically determined dental development against the metrics of the *pars basilaris*, humeri and femoral growth as the standard for developmental age for this sample.

2. Materials and Methods

The Poulton Research Project is a rural, multi-period site in Poulton, Cheshire, United Kingdom. Since 1995, research and excavations have provided information about the Mesolithic, Neolithic, Bronze and Iron Ages, as well as the Roman and Medieval periods (Emery, 2000).

One of the current focuses of the project is the excavation of the medieval chapel and its surrounding graveyard, which was used by the local population from approximately 1150 to 1493 (Emery, 2000). Through to the end of the 2013 excavation season, the full outline of the chapel had been revealed exposing the foundations of a tower, nave and chancel. The cemetery population

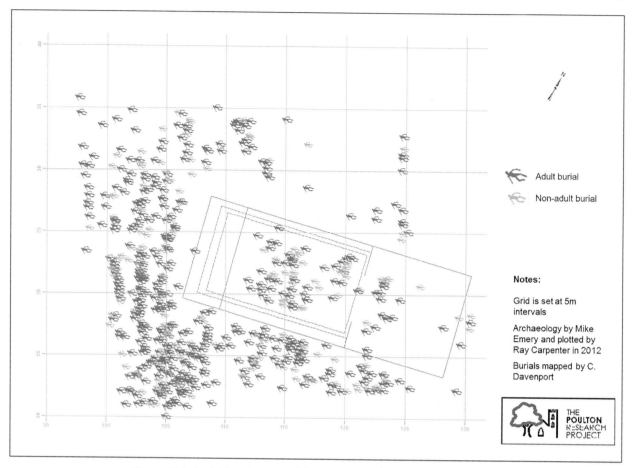

Adult burial

Non-adult burial

Notes:

Grid is set at 5m intervals

Archaeology by Mike Emery and plotted by Ray Carpenter in 2012

Burials mapped by C. Davenport

THE POULTON RESEARCH PROJECT

Figure 10.1 Burial location map of the medieval chapel at Poulton, Cheshire.

initially comprised of the lay community for the Abbey of St Mary and St Benedictine, a Cistercian abbey and grange founded by Robert the Butler between 1146 and 1153 (Fisher, 1984). It was during this initial construction phase that the initial single cell structure of Poulton Chapel was founded, most likely to provide a temporary place of worship for the monks. Following the movement of the Cistercians to Dieulacres Abbey, Leek, Staffordshire in 1214, the remaining lay brothers (*conversi*) and local population continued to use the chapel until it became a private chapel in 1493 (Emery, 2000). A total of 757 human burials had been identified and excavated by the end of the 2013 excavation season, of which 593 were articulated skeletons. It is likely that the graveyard may contain upwards of 1500 burials.

Most burials were outside the chapel walls, particularly on its south and west sides, with some burials recovered from inside the nave and chancel (Figure 10.1). Nearly all human remains were buried with an east–west orientation (head facing east), as is typical of Christian burials (Daniell, 1998). There is no apparent spatial distribution segregation between adult and non-adult (<18 years at death) burials, but there is one non-adult cluster southwest of the chapel (Figure 10.1). However, excavations on the north and east sides of the chapel have been limited so far, which may have skewed the information available.

The sample here was drawn from the 244 non-adult skeletons recovered through to the end of the 2013 season. Sufficient dental remains were preserved from 171 non-adult skeletons to permit their seriation into developmental order. This seriation utilised maxillae and/or mandibles, and included radiographs of each dentition. Dental radiographs were obtained using a Hewlett-Packard Faxitron model 43855A (Faxitron Bioptics, LLC, 3440 East Britannia Drive, Suite 150, Tucson, Arizona 85706 USA) with the Digital Linear X-Ray Scanner EZ240 and iX-Pect for EZ240 Software (NTB GmbH, Doenseler Strasse 110, D 49453 Dickel, Germany). The radiographs were calibrated at 60kv, 100ms – 110kv, 100ms, dependent on the development of the individual, then processed using MicroDicom 0.8.6 analysis software (MicroDicom, Sofia, Bulgaria). Images were taken mediolaterally, and when required, anteroposteriorly. Dental development was evaluated using Ubelaker (1999), Moorrees (1963a, b) and AlQahtani *et al.* (2010). Special care was taken when anomalies were present (e.g., congenital absence of the third molars; in rare cases congenital absence of the lateral incisors). Figure 10.2 shows an example of dental development in Poulton non-adults.

Table 10.1 shows the sample sizes for those skeletons in the dental seriation that also preserved humeri and/or femora (Figure 10.3). A sample of 55 individuals in

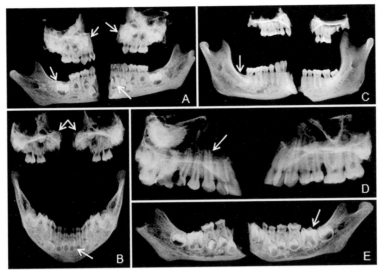

Figure 10.2 Dental radiographs showing: (A and B) calcification of the permanent dentition within the alveolar bone; (C) asymmetrical congenital absence of the third permanent molar; (D) incomplete root formation; (E) eruption of the first permanent molar.

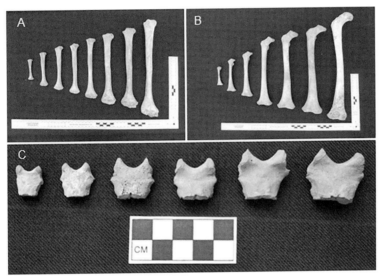

Figure 10.3 (A) Selection of seriated right humeri; (B) right femora; (C) pars basilaris.

Table 10.1 Samples of subadult humeri and femora.

	Both left and right	Left only	Right only
Humeri	51	20	27
Femora	74	14	16

the dental seriation also had a preserved *pars basilaris* (Figure 10.3).

Once seriated into developmental order, age markers were placed into the seriation that coincided with calcification and eruption stages presented by AlQahtani *et al.* (2010). Other dental methods were not used for this step to enable consistency between the markers added. As the London Atlas of Tooth Development and Eruption had been developed, in part, using both the Spitalfields collection and radiographs from living individuals, it would also ensure the greatest consistency between the ages assigned to the seriation and the comparison populations.

Maximum diaphyseal lengths were obtained for each humerus and each femur using an osteometric board (Figure 10.4). Maximum width (MWB), maximum length (MLB) and sagittal length (SL) were obtained for each *pars basilaris* using sliding a calliper (Figure 10.4). Most individuals had both their left and right humeri and/or their left and right femora available for analysis (Table 10.1). For those individuals that preserved both sides, left and right humeri did not differ significantly in length. Similarly, left and right femora also did not differ significantly in length. Nevertheless, left and right diaphyseal lengths were analysed separately to increase sample sizes by inclusion of those for which only one side was recovered.

When assessing dental and skeletal samples, any individuals that exhibited signs of chronic disease, stress, trauma or developmental defects were excluded from this study due to the potential effects on growth patterns (Mays, 1995; Ribot and Roberts, 1996; Steckel, 2005).

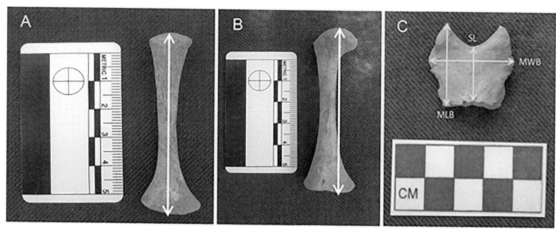

Figure 10.4 Metrics: (A) maximum humeral diaphyseal long bone length; (B) maximum femoral diaphyseal long bone length; (C) the three measurements (MWB, MLB & SL) obtained from each pars basilaris.

Although it was necessary to exclude some individuals, this number was low (n=3) suggesting a fairly healthy and robust population.

Only two individuals were excluded during the measurements of the humeri and femora for this sample. The first is a 3–4 year old individual presenting distinct skeletal changes of rickets in the bones of both legs, each illustrating the classical bending deformities (Ortner, 2003). The other individual demonstrates a healed greenstick facture to the left tibia (Lovell, 1997). This trauma, although healed and well aligned, has led to a shortened length in comparison to the 'healthy' side. As a result, the growth and length of the associated femora have been affected, leaving the left femora longer in comparison to the right side. The authors highlight that further research is currently underway to explore other indicators of stress within this collection. Indicators currently under exploration include porotic hyperostosis, linear enamel hypoplasia, cribia orbitalia and Harris lines of arrested growth. Preliminary results suggest that the frequency of these indicators is low for this collection (Dove et al., 2016).

A further individual was eliminated from the pars basilaris study due to a developmental defect. A cleft pars, or bifid clivus, highlights a condition where the midline of the pars basilaris fails to integrate with the axial sclerotome during development, leading to a retained split anterior to the facets for the jugular and condylar limbs of the pars lateralis (Menezes, 1997; Müller and O'Rahilly, 2003; Pang and Thompson, 2011). Dependent on severity, this split can extend up to the spheno-occipital synchondrosis, leading to the element presenting in multiple parts (Pang and Thompson, 2011). In this study, the specimen excluded from the sample exhibited a partial cleft deformity, leading to the bone presenting as a single element, but with changes considered significant enough to bias the sample.

3. Results

Nonparametric correlations (Kendall's T and Spearman's ρ) for all metrics with seriated dental development rank order were very high and significant

Table 10.2 Nonparametric correlations comparing diaphyseal lengths and pars basilaris metrics with relative physiological age based on radiographs seriated into developmental order.

Maximum diaphyseal length	N	Kendall's T	Spearman's ρ
Left humerus	71	0.808***	0.928***
Right humerus	78	0.882***	0.973***
Left femur	88	0.860***	0.965***
Right femur	90	0.880***	0.975***
Pars Basilaris			
Maximum length	51	0.716***	0.878***
Maximum width	51	0.700***	0.873***
Sagittal length	52	0.619***	0.805***

*** $P < 0.001$

(for all cases $P < 0.001$; Table 10.2). Nonparametric correlations for maximum humeral and femoral diaphyseal lengths were consistently higher than the three pars basilaris measurements. The weakest, though still highly significant correlation ($P < 0.001$), was for sagittal length of the pars basilaris.

Relative to seriated dental development rank order, growth of the long bones and the pars basilaris growth are shown in Figures 10.5 and 10.6, respectively. In both figures, dental developmental markers are identified to provide chronological age equivalent landmarks against which seriated dental developmental rank order can be evaluated. These landmarks are the eruptions of the first deciduous molar (DM1), the second deciduous molar (DM2), the first permanent molar (M1), second permanent molar (M2) and the third permanent molar (M3).

In general, long bone growth is curvilinear (Figure 10.5). Relative to the humeri, femoral diaphyseal growth is more rapid early up until about the eruption of DM2, at approximately 2 years of age (AlQahtani et al., 2010). Thereafter, humeral and femoral growth rates are similar, with both accelerating with M2 eruption. This corresponds with the time of pubescence, which has been estimated

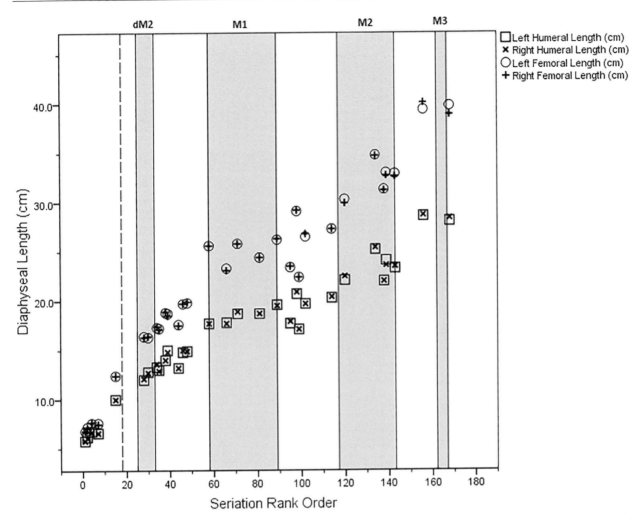

Figure 10.5 Scatterplot demonstrating the curvilinear relationship of each of the humeral and femoral diaphyseal maximum lengths versus seriated dental developmental rank order. Dental development bands (dM2, M1, M2 and M3) are applied to enable age to be ascertained from the seriation rank order. Dashed line denotes commencement of eruption of first deciduous tooth.

to occur between 11 and 12 years of age (Shapland and Lewis, 2013).

Growth of the *pars basilaris* is curvilinear and slows with increasing age (Figure 10.6). Growth in sagittal length occurs early and increases only slightly after eruption of DM1 at approximately 12–18 months of age (AlQahtani *et al.*, 2010). Growth in maximum length and width is accelerated up until the eruption of DM2. Growth in both of these latter dimensions is much more rapid than for sagittal length. Coinciding with the M1 eruption at approximately 6 years of age (AlQahtani *et al.*, 2010), the growth rates for the *pars basilaris* slows considerably with decreased rates of growth noted in all measurements.

4. Discussion

Relative to known age, Maresh (1970) reported maximum humeral and femoral diaphyseal lengths obtained from radiographs of a sample of living non-adults (Figure 10.7) from Denver, Colorado, USA. Her sample consisted of 255 subjects from birth through to 19 years of age (1933–1967). These subjects were considered to

be of northwest European decent and this study was recognised as a worthy longitudinal study by the World Health Organisation (McCammon, 1970). Maresh (1970) reported these measurements relative to chronological age (i.e., calibrated with time since birth) for these living non-adults, which differs from the dental development rank order (i.e., essentially calibrated with time since fertilisation) used here for the Poulton non-adults. Nonetheless, growth patterns are remarkably similar between these two populations (see Figures 10.5 and 10.7). This similarity provides strong support for the use of seriated dental developmental rank order as a good proxy for relative age.

For estimating chronological age at death using the *pars basilaris*, Fazekas and Kósa (1978) provided data for antenatal individuals, while Redfield (1970) provided descriptions for non-adults. Therefore, these reports are not directly comparable because the former (Fazekas and Kósa, 1978) provides antenatal data and the latter (Redfield 1970) is qualitative.

Regardless, growth patterns are broadly similar between the results reported by Scheuer and MacLaughlin-Black

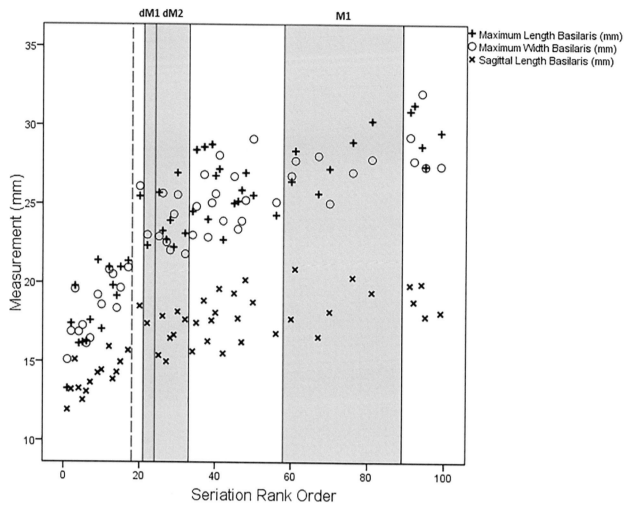

Figure 10.6 Scatterplot demonstrating the curvilinear relationship of each of the three pars basilaris *measurements versus seriated dental developmental rank order. Dental development bands (dM1, dM2 and M1) are applied to enable age to be ascertained from the seriation rank order. Dashed line denotes commencement of eruption of first deciduous tooth.*

(1994) and the Poulton non-adults (see Figures 10.6 and 10.8). As with the humeri and femora, this similarity provides strong support for the use of seriated dental developmental rank order as a good proxy for relative age.

Scheuer and MacLaughlin-Black (1994) provided quantitative data for the *pars basilaris* for a sample of 46 non-adults, drawn from the Spitalfields (Natural History Museum, London) and the St Bride's Collections (Museum of London). Molleson and Cox (1993) also reported *pars basilaris* metrics for 28 non-adults from the Spitalfields Collection, each identified by a catalogue number. Figure 10.8 demonstrates that at least some of the individuals reported by Molleson and Cox (1993) were also included in the sample used by Scheuer and MacLaughlin-Black (1994). Therefore, this duplication of data does serves to limit the total number of available samples from which age can be assessed. When the sample is broken down into age categories, the number of individuals representing each specific age category can reduce to sample sizes of less than four (Scheuer and MacLaughlin-Black, 1994). For example, the total sample reported was spread over individuals whose age at death ranged from 2 weeks to 4

years 7 months, meaning the majority of age categories contained only a single individual.

The small sample sizes have also produced a tendency for the methods from Scheuer and MacLaughlin-Black (1994) to consistently overage the Poulton non-adults, placing them in higher age categories than with other age assessments. In addition, both Redfield (1970) and Scheuer and MacLaughlin-Black (1994) noted that *par basilaris* maximum length exceeds maximum width until five months postpartum, and reverses thereafter. However, this is not true for the Poulton non-adult sample, where length can exceed width at all ages (Figure 10.8). This is most likely due to the comparative sample sizes used in these studies not fully representing the variability within any given chronological age, as the differences in developmental age are not taken into account.

One approach is to improve non-adult sample sizes. By minimising the inherent small sample size biases by removing the implicit restriction that only samples with known chronological age at death are suitable for growth studies and the establishment of standards will help address this issue. Developmental age is as relevant

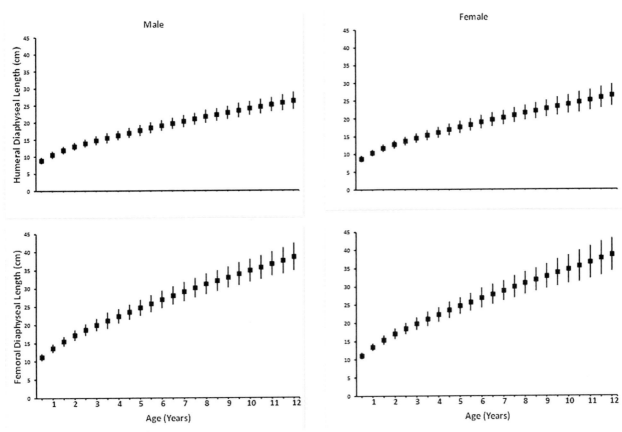

Figure 10.7 Humeral and femoral diaphyseal length versus chronological age in half year intervals. Data were obtained from Maresh (1970). Means are shown with error bars displaying ± 2 standard deviations.

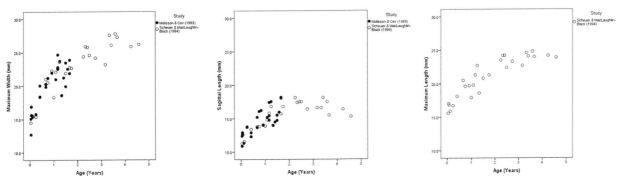

Figure 10.8 Maximum length, sagittal length and maximum length of the pars basilaris versus chronological age at death. Data were obtained from Molleson and Cox (1993) and Scheuer and MacLaughlin-Black (1994).

to growth studies as is chronological age, and even eliminates the 'arbitrary zero calibration point' of birth.

Large non-adult samples, such as the growing collection recovered from the Medieval Poulton chapel graveyard, offer an alternative. Although their chronological ages are not known, these can be assessed using developmental age. Seriation offers a viable method for placing a non-adult sample in rank order based on developmental age. The use of seriation provides the ordinal data against which growth studies can be analysed. The ultimate goal will be to calibrate developmental age rank order seriations against chronological ages.

Concerns surrounding the living conditions within archaeological periods, with frequent childhood illnesses and malnutrition, could lead to disrupted growth and development within a sample. Individuals affected by chronic disease or malnutrition may present puberty delay (Lewis *et al.*, 2016; Primeau *et al.*, 2016). However, an acute illness would essentially cause a swift death leaving no apparent scar to their skeleton. Therefore, for an illness to affect and become osteologically evident, an individual must be subjected to prolonged exposure to illness or malnutrition (Lovejoy *et al.*, 1990). For this sample, some non-adult individuals do present skeletal stress indicators. However, these individuals were removed and not included in the analysed sample. Respectfully, it could be argued that the growth of these long bones are not representative of this sample as these non-adults have died before reaching adulthood. This statement has been tested and identified that this effect is too small to be significant when compared to other methodological considerations (Primeau *et al.*, 2016).

5. Conclusions

There is marked similarity between Poulton and known age at death populations in growth patterns of the humerus, femur and the *pars basilaris* (see Figures 10.5–10.8). The initial research presented here is being expanded for these and additional skeletal elements, which will eventually provide growth standards for the Poulton non-adults against which other populations can be compared. Furthermore, the individuals that were excluded from this study due to exhibiting signs of stress or developmental defects are undergoing further analysis to determine the impact that stress in childhood had on the growth patterns observed in the Poulton sample.

Prior to assessing the biological age of individuals from a population, it is recommended that the comparative known age at death sample contains a statistically viable number of individuals at each age range to account for variability in growth due to lifestyle or environment. This study has demonstrated that the dental developmental age seriation obtained for the Poulton non-adult sample recovered so far shows great promise as a technique to enable the evaluation of biological age estimation.

Acknowledgements

Our special thanks go to Michael M. Emery and the Trust from the Poulton Research Project for their full support and allowing the study of the human remains collection. We wish to thank Dr Jessica Pearson for allowing access to the portion of the Poulton Collection currently housed at the University of Liverpool. Finally, we wish to thank Colin Armstrong, from Liverpool John Moores University, for his laboratory assistance.

References

AlQahtani, S.J., Hector, M.P., Liversidge, H.M. 2010. Brief communication: The London atlas of human tooth development and eruption. *American Journal of Physical Anthropology* 142(3), 481–490.

Anderson, M., Messner, M.B., Green, W.T. 1964. Distribution of lengths of the normal femur and tibia in children from one to eighteen years of age. *Journal of Bone and Joint Surgery* 46A(6), 1197–1202.

Belsky, D.W., Caspi, A., Houts, R., Cohen, H.J., Corcoran, D.L., Danese, A., Harrington, H., Israel, S., Levine, M.E., Schaefer, J.D., Sugden, K. 2015. Quantification of biological aging in young adults. *Proceedings of the National Academy of Sciences* 112(30), 4104–4110.

Cameron, N. Demerath, E.W. 2002. Critical periods in human growth and their relationship to diseases of aging. *Yearbook of Physical Anthropology* 45 (Supplement 35), 159–184.

Cardoso, H.F.V. 2006. Brief communication: The collection of identified human skeletons housed at the Bocage Museum (National Museum of Natural History), Lisbon, Portugal. *American Journal of Physical Anthropology* 129(2), 173–176.

Daniell, C. 1998. *Death and Burial in Medieval England, 1066–1550*. Routledge, London.

Demirjian, A., Goldstein, H., Tanner, J.M. 1973. A new system of dental age assessment. Human *Biology* 45(2), 211–227.

Dove, E.R., Irish, J., Eliopoulos, C., De Groote, I. 2016. Re-evaluating the co-occurrence and age of formation of Harris lines and linear enamel hypoplasia. *American Journal of Physical Anthropology* 159, 133–133.

Emery, M.M. 2000. *The Poulton Chronicles: Tales from a Medieval Chapel*. Poulton Archaeology Press, Williamsburg, VA.

Fazekas, I.G., Kósa, F. 1978. *Forensic Fetal Osteology*. Akadémiai Kiadó, Budapest.

Fisher, M.J. 1984. *Dieulacres Abbey. Leek Staffordshire*. Leek, Dieulacres Abbey, 2nd Edition.

Hoffman, J.M. 1979. Age estimations from diaphyseal lengths: Two months to twelve years. *Journal of Forensic Science* 24(2), 461–469.

Jukic, A.M., Baird, D.D., Weinberg, C.R., McConnaughey, D.R., Wilcox, A.J. 2013. Length of human pregnancy and contributors to its natural variation. *Human Reproduction* 28(10), 2848–2855.

Lampl, M. Johnston, F.E. 1996. Problems in the aging of skeletal non-adults: Perspectives from maturation assessments of living children. *American Journal of Physical Anthropology* 101(3), 345–355.

Lampl, M., Veldhius, J.D., Johnson, M.L. 1992. Saltation and stasis: A model of human growth. *Science* 258, 801-803.

Langley-Evans, S.C. 2015. Nutrition in early life and the programming of adult disease: A review. *Journal of Human Nutrition and Dietetics* 28(1), 1–14.

Lewis, A.B., Garn, S.M. 1960. The relationship between tooth formation and other maturational factors. *The Angle Orthodontist* 30(2), 70–77.

Lewis, M.E. 2007. *The Bioarchaeology of Children*. Cambridge, Cambridge University Press.

Lewis, M., Shapland, F., Watts, R. 2016. The influence of chronic conditions and the environment on pubertal development. An example from medieval England. *International Journal of Paleopathology* 12, 1–10.

Lovejoy, C.O., Russel, K.F., Harrison, M.L. 1990. Long bone growth velocity in the Libben population. *American Journal of Human Biology* 2, 533–541.

Lovell, N.C. 1997. Trauma analysis in paleopathology. *Yearbook of Physical Anthropology* 40, 139–170.

Maresh, M.M. 1970. Measurements from roentgenograms, heart size, long bone lengths, bone, muscle and fat widths, skeletal maturation, in: McCammon, R.W. (Ed.), *Human Growth and Development*. Charles C. Thomas, Springfield, IL, pp. 155–200.

Mays, S. 1995. The relationship between Harris lines and other aspects of skeletal development in adults and juveniles. *Journal of Archaeological Science* 22, 511–520.

Menezes, A.H. 1997. Craniovertebral junction anomalies: Diagnosis and management. *Seminars in Pediatric Neurology* 4(3), 209–223.

Milne, G. 1997. *St Bride's Church London Archaeological Research 1952–60 and 1992–5*. English Heritage, London.

Molleson, T. 1990. The children from Christ Church crypt, Spitalfields. *American Journal of Physical Anthropology* 81(2), 271.

Molleson, T.I., Cox, M. 1993. *Spitalfields Project: Volume 2: The Anthropology - The Middling Sort.* Council for British Archaeology, London.

Moorrees, C.F.A., Fanning, E.A., Hunt, E.E. 1963a. Formation and resorption of three deciduous teeth in children. *American Journal of Physical Anthropology* 21, 205–213.

Moorrees, C.F.A., Fanning, E.A., Hunt, E.E. 1963b. Age variation of the formation stages for ten permanent teeth. *Journal of Dental Research* 42(5), 1490–1501.

Müller, F., O'Rahilly, R. 2003. Segmentation in staged human embryos: The occipitocervical region revisited. *Journal of Anatomy* 203(3), 297–315.

Ortner, D.J. 2003. *Identification of Pathological Conditions in Human Skeletal Remains.* London, Academic Press.

Pang, D., Thompson, D.N.P. 2011. Embryology and bony malformations of the craniovertebral junction. *Child's Nervous System* 27(4), 523–564.

Primeau, C., Friis, L., Sejrsen, B., Lynnerup, N. 2012. A method for estimating age of Danish Medieval sub-adults based on long bone length. *Anthropologischer Anzeiger* 69, 317–333.

Primeau, C., Friis, L., Sejrsen, B., Lynnerup, N. 2016. A method for estimating age of Medieval sub-adults from infancy to adulthood based on long bone length. *American Journal of Physical Anthropology* 159(1), 135–145.

Redfield, A. 1970. A new aid to aging immature skeletons: Development of the occipital bone. *American Journal of Physical Anthropology* 33(2), 207–220.

Ribot, I., Roberts, C. 1996. A study of non-specific stress indicators and skeletal growth in two mediaeval subadult populations. *Journal of Archaeological Science* 23(1), 67–79.

Rissech, C., Schaefer, M., Malgosa, A. 2008. Development of the femur—implications for age and sex determination. *Forensic Science International* 180(1), 1–9.

Scheuer, L., Black, S. 2000. *Developmental Juvenile Osteology.* Academic Press, London.

Scheuer, L., MacLaughlin-Black, S. 1994. Age estimation from the pars basilaris of the fetal and juvenile occipital bone. *International Journal of Osteoarchaeology* 4, 377–380.

Schour, I., Massler, M. 1941. The development of the human dentition. *Journal of the American Dental Association* 28, 1153–1160.

Shapland, F., Lewis, M.E. 2013. Brief communication: A proposed osteological method for the estimation of pubertal stage in human skeletal remains. *American Journal of Physical Anthropology* 151(2), 302–310.

Steckel, R.H. 2005. Young adult mortality following severe physiological stress in childhood: Skeletal evidence. *Economics and Human Biology* 3(2), 314–328.

Stull, K.E., L'Abbé, E.N., Ousley, S.D. 2014. Using multivariate adaptive regression splines to estimate subadult age from diaphyseal dimensions. *American Journal of Physical Anthropology* 154(3), 376–386.

Ubelaker, D.H. 1987. Estimating age at death from immature human skeletons: An overview. *Journal of Forensic Sciences* 32(5), 1254–1263.

Ubelaker, D.H. 1999. *Human Skeletal Remains: Excavation, Analysis and Interpretation.* Taraxacum, Washington, D.C.

Willey, P., Galloway, A., Snyder, L. 1997. Bone mineral density and survival of elements and element portions in the bones of the Crow Creek massacre victims. *American Journal of Physical Anthropology* 104(4), 513–28.